HANDBOOK OF CONTROL ROOM DESIGN AND ERGONOMICS

A Perspective for the Future

Second Edition

HANDBOOK OF CONTROL ROOM DESIGN AND ERGONOMICS

A Perspective for the Future
Second Edition

Edited by

TONI IVERGÅRD AND BRIAN HUNT

CRC Press
Taylor & Francis Group
Boca Raton London New York

CRC Press is an imprint of the
Taylor & Francis Group, an **informa** business

CRC Press
Taylor & Francis Group
6000 Broken Sound Parkway NW, Suite 300
Boca Raton, FL 33487-2742

First issued in paperback 2019

© 2009 by Taylor & Francis Group, LLC
CRC Press is an imprint of Taylor & Francis Group, an Informa business

No claim to original U.S. Government works

ISBN-13: 978-1-4200-6429-2 (hbk)
ISBN-13: 978-0-367-38673-3 (pbk)

Library of Congress Cataloging-in-Publication Data

Handbook of control room design and ergonomics : a perspective for the future / editors, Toni Ivergård [and] Brian Hunt. -- 2nd ed.
   p. cm.
   Prev. ed. cataloged under author Toni Ivergård.
   Includes bibliographical references and index.
   ISBN 978-1-4200-6429-2 (hardback : alk. paper)
   1. Human engineering--Handbooks, manuals, etc. 2. Control rooms--Design and construction--Handbooks, manuals, etc. I. Ivergård, Toni. II. Hunt, Brian. III. Ivergård, Toni. Handbook of control room design and ergonomics.

TA166.I94 2009
620.8'2--dc22   2008026813

Visit the Taylor & Francis Web site at
http://www.taylorandfrancis.com

and the CRC Press Web site at
http://www.crcpress.com

Contents

PART I   Introduction: Work in Control Rooms and Models of Control Room Work

PART II   Design of Information and Control Devices

PART III   Design of Control Rooms and Their Environment

PART IV   Case Studies and Applications

PART V   The Human Dimension in the Control Room

PART VI   Conclusions

Chapter 14
Toni Ivergård and Brian Hunt

Preface to the First Edition

A truck with its driver on board is standing on a sloping quayside. Suddenly the truck starts to roll backward toward the edge of the quay. The driver reacts spontaneously and pushes the direction lever forward. This results in the truck increasing speed and falling over the edge of the quay. The truck driver is seriously injured in the accident.

In the police enquiry following this incident, it was concluded that the accident was due to the truck driver's error. However, the lever that controlled the direction in which the truck travelled was designed so that when the lever was pushed forward the truck travelled backward, and when it was pushed backward, the truck moved forward.

On further thought, it is self-evident that this design is both illogical and unsuitable. It should instead have been designed so that the truck travelled in the direction of the lever movement. It could be said that the accident was caused by designer error rather than user error. The designer had not taken into account the natural response of a truck driver in such an incident—that is to say—no consideration had been given to ergonomic factors in the design of the controls of the truck.

Ergonomics is concerned with the adaptation of technology to suit man's natural abilities and needs. The aims of research in this field are to:

1. Prevent ill-health, injury, and fatigue.
2. Create comfort and well-being.
3. Promote efficiency (both qualitatively and quantitatively) and improve reliability in production.
4. Create interesting and meaningful jobs.

Control rooms of different types have considerable problems as far as adaptation to man is concerned. Problems include instruments that are difficult to read and which are illogically positioned, or controls that are difficult to hold and have unsuitable movement directions.

This handbook is designed to be used as a simple and easily accessible aid in the design and development of new control and operation rooms, primarily in the process industries. Suitable sections may also be used in the design of other types of control rooms. The book is important for all those involved in the design of operation: control room designers, instrument engineers, data engineers, process specialists, and control engineers. It is my hope that the process operators who sit and operate the machines will also be able to use this book.

It is sometimes impossible to provide concrete recommendations and solutions to problems. What is needed instead is a general knowledge of man's functioning, needs, and nature in order to be able to choose the best technical solution in a particular situation. This results in certain chapters in the book being of a more descriptive nature.

Accordingly, this could be seen as an ergonomic 'cookbook' where the 'recipes'—that is, the development of operation and control rooms—are easily accessible.

A lot of background material and many definitions have been excluded in order to fulfil this aim, but those who are especially interested can find them in other publications cited.

Toni Ivergård
Stockholm, Sweden

Preface to the Second Edition

This *Handbook of Control Room Design and Ergonomics* was first published in 1989 by Taylor & Francis. The first edition, produced nearly 20 years ago, was based on rather classic approaches to ergonomics, as was the concept of design. However, in spite of this we noticed that this book, although long out of print, was still selling surprisingly well on the secondhand market. The first edition was based on solid practical research on control room work from most areas of industry. This information is still relevant and has inspired us to produce this new, extensively updated edition.

This new edition has two functions. First, it covers more extensively the use of the control room and its related computer system beyond the traditional tasks of process monitoring and supervision. Second, it describes the use of the control system for optimising and developing the existing systems and processes. The control room can also be used for the purposes of education, learning, and simulation training. But at the utmost, the control room of the future should become a high-ceiling environment for creativity and innovation. By 'high ceiling' we mean an environment that has a high tolerance for error and is thus a suitable environment for learning from error. Over the past year, we have jointly and separately been researching in areas of relevance with a view to updating and rewriting this classic handbook. This second edition still aims to be a practical handbook of guidelines and cases.

The concept of 'control rooms' has changed and expanded enormously during the time between the first and second editions. In composing this new edition, we have catered to these developments. Accordingly, this new edition has a section describing this new situation and also includes a new taxonomy/paradigm. Nowadays, as their roles and functions have greatly expanded, it is often more appropriate to talk about control centres. A modern approach for looking at work in control rooms uses recent concepts of creativity and learning/developing environments. We have incorporated such concepts into the current text. Additionally, we have included new ideas and philosophies about organisational design and job design as these are applied to control room-related work. We therefore include and describe some creative organisational designs of the future. Learning organisations and learning at work are integral parts that utilise the information and communication technology (ICT) potential of modern control systems. In this respect, we have added some theoretical background about learning, learning in the workplace, and lifelong learning. Today, process control encompasses a new generation of computer systems that have enormous capabilities, including the potentials of new display technologies. In other words, advanced technologies are today, to a very large extent, integrated and interrelated with human factors and organisational development.

We consider some basics of ergonomics of controls and displays to be very important and for this reason we have retained these concepts and models. But we have added a major part related to all new innovations in large-scale information displays. These new features also influence the design and layout of the control room. They will also help the reader develop a better understanding and insight, particularly for relevant creative work. In the course of writing this second edition, we approached leading-edge companies for ideas and suggestions, and they generously assisted us in providing a number of exciting design examples. We are grateful for the helpful insights from these colleagues from industry.

Concepts of knowledge management, data mining, and artificial intelligence (AI), including the use of logistics, queuing theory, and so forth, also have high potential in the creation of the control processes and control rooms of the future. Ideally, control room work of the future is no longer a tedious or boring, monotonous task solely focused on low vigilance of work processes. It is now a stimulating creative design for optimising system performance and shaping the future as part of business development efforts to improve competitiveness while conserving scarce resources and saving the environment.

Toni Ivergård
Brian Hunt
Bangkok, Thailand

Acknowledgments

Jatupol Chawapatnakul redrew and developed most of the artwork for this new edition. He also transferred the text and illustrations of the old book into a software format for the new edition.

The Swedish power company Vattenfall has over decades supported research into the human factor aspects of the design of control systems, safety and general information about power production, and related distribution systems. The company generously shared information with the authors about their new business in energy and carbon trading.

Eyevis GmbH, Germany, and dnp Denmark contributed a large number of illustrations and design information of new display technology.

Stora Enso contributed photos as well as shared information about their new approach to control room design.

Wallenins Marine also provided photographs and information.

Editors

Toni Ivergård, Ph.D., is Director of the Master of Management in Innovation and Entrepreneurship at Rangsit University, Bangkok, Thailand. He is also Chairman of the Board and Managing Director of Entrepreneurship Swedish Thai Co. Ltd.

Dr. Ivergård was the author of the first edition of this book. He and Brian Hunt have edited and rewritten all parts of the current edition and added new chapters and case studies containing practical examples.

Over a 10-year period Dr. Ivergård conducted research and consultancy in the area of control room design in most areas of application, for example power production and distribution, shipping, paper and pulp industries, airlines, and traffic control systems. He attained his Ph.D. from Loughborough University, United Kingdom. Dr. Ivergård is the author of a number of journal articles and official reports on public sector reform. His research interests are learning in organisations, public sector reforms, and ergonomic approaches to workplace learning.

Brian Hunt is Assistant Professor at the College of Management at Mahidol University, Bangkok, Thailand. He is the coauthor and editor with Toni Ivergård of this handbook.

Hunt has conducted research into learning pedagogy and learning at work, and has a number of academic publications in this area. He is the author of a number of journal articles on learning in organizations and public sector reform. His research interests are learning in organizations, public sector organisational behaviour and development, and management in cross-cultural environments.

Contributors

Erik Dahlquist
Mälardalen University
Vasteras, Sweden

Eric Hénique
Eyevis GmbH
Reutlingen, Germany

Soeren Lindegaard
dnp Denmark
Karlslunde, Denmark

Monica Lundh
Chalmers University of
  Technology
Gothenburg, Sweden

Margareta Lützhöft
Chalmers University of
  Technology
Gothenburg, Sweden

Andy Nicholl
Consultant in Occupational
  Hygiene and Acoustics
Dunfermline, Scotland

Part I

Introduction: Work in Control Rooms and Models of Control Room Work

1 Work in Control Rooms

Toni Ivergård and Brian Hunt

CONTENTS

Traditionally, human work was of the handicraft type engaged in by workers using only the very simplest of tools. Normally, whole handicraft items were produced by the work of one person from beginning to end. This traditional approach to the organisation of the work process began to change during the industrial revolution. Industrialisation brought large and complex machines into the workplace, though these were initially only used to augment the available power—workers were still operating manufacturing machinery by hand. Perhaps the largest change occurred in the organisation of work. Instead of a single worker making the whole or a major part of a product, the task was broken down into several smaller processes each carried out by a different worker. Figure 1.1 shows the development of the human role in the different stages of automation. In Figure 1.1a there is direct human control without automation, whilst in Figure 1.1b a low level of automation has been introduced. In Figure 1.1c a control computer is used to coordinate the different local control devices, and Figure 1.1d is the most advanced control system, based on integrated communication between a large number of small computers.

Technology changed more quickly in some industries than in others. It soon became possible to use the available technology to convert raw materials into the final product with minimal direct contact with the worker. A technical process had taken over the former manual manufacturing process of the product.*

---

* At the time of the first edition of this handbook, the concept of a process industry encompassed a whole variety of industries and companies that all had one feature in common. In these industries, the input (raw material) is processed, converted, or changed (often chemically and/or physically) in order to create the finished product. In process industries, it is also common for subprocesses to occur in some sort of flow, and for the progressions between the different elements to be more or less directly connected to each other. These subprocesses are often automated; that is, they occur without any direct intervention from a worker. Typical examples of process industries are the paper, electricity, food, gas, and petroleum industries. In all these industries the raw material and the finished product are very different, and although the work carried out by people in each system varies markedly, the principles of the organisation of work processes are the same. Since the first edition, there has been an explosion in information handling in all areas of business and society. Naturally, this has had implications for understanding the concept of process industries. The markets for trading of energy, natural resources, and environmental pollution are all, in a manner of speaking, kinds of process industries. The use of computers as well as information and computer technologies (ICT) is an important factor that adds on to this greatly expanded conceptualisation of process industries.

3

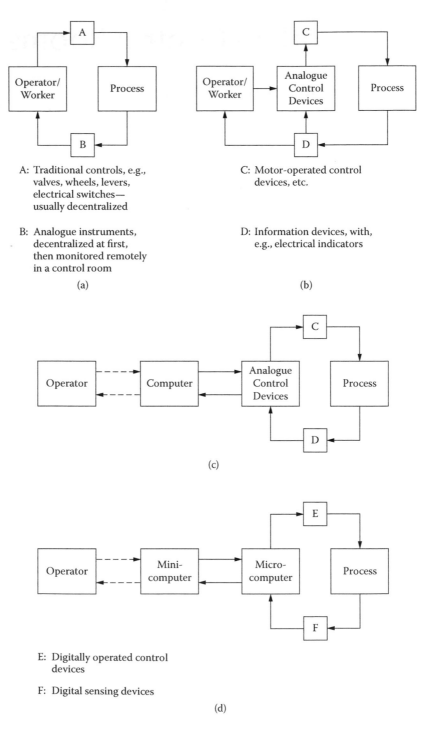

A: Traditional controls, e.g., valves, wheels, levers, electrical switches— usually decentralized

B: Analogue instruments, decentralized at first, then monitored remotely in a control room

(a)

C: Motor-operated control devices, etc.

D: Information devices, with, e.g., electrical indicators

(b)

(c)

E: Digitally operated control devices

F: Digital sensing devices

(d)

FIGURE 1.1   Different levels of automation and the change in the human role from online controller to off-line supervisor and planner.

In a process industry the operator is often responsible for expensive equipment and for the health and safety of the workers. Some have found that these responsibilities lead to stress and hesitancy on the part of the operator (Johansson, 1982).

1.1   COMPUTERISATION

Computerisation often involves the amalgamation of the larger process units into a single control room. This adds to the operator's responsibilities since the economic value of what is controlled increases, which in turn increases the fear of any fault arising. In addition, the operator ends up further away from the actual process. There is thus a likelihood that the operator's knowledge will diminish about what actually happens in the process (that is, the opportunity for learning and improvement also diminishes). Over time, this can mean that the operator is uncertain of what to do in critical situations, thereby heightening the feelings of fear and insecurity.

It is interesting to note that, contrary to expectations, computerisation in the process industries has not brought any great reduction in the number of employees. According to Berglund (1982), a limited level of automation may result in a certain reduction in staffing levels (see Figure 1.2). But when the degree of automation rises, a 'cut-off' level is eventually reached where there is no further reduction in manning levels. If the degree of automation is increased much further, the staffing level may even increase. Sorge et al. (1982) found similar results in the manufacturing industry.

The introduction of computers for process control and process monitoring brings with it new forms of information and control devices. The traditional control panel has been either partially or completely replaced by visual display units (VDUs), and the traditional buttons, levers, and knobs have been replaced by keyboards. This can create new problems. VDUs require special lighting conditions and the screens increase the load on the operator's eyesight. The various types of printers can create considerable noise problems. Certain units give off a relatively large amount of heat. For example, a data terminal gives off about the same amount of heat as a person. If the ventilation system is already working near its limit, problems of excessive warmth may occur. This list of problems can be extensive. Here, however, it suffices to say that there are a variety of different problems that may be encountered in

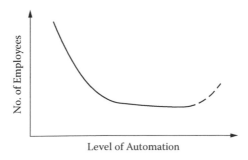

FIGURE 1.2   The relation between level of automation and number of employees.

computerisation of process control. It is therefore important to attempt to predict the problems at an early stage in the planning of a new, computerised process industry (or in the computerisation of an older existing process industry) so that they can be avoided as far as possible.

Figure 1.3 shows schematically the progression towards a technical system. The initial stages involve the planning and design, and then the system is built. This is operated and maintained, and at the end is finally dismantled and partially or completely removed. Traditionally, planning has been carried out as in Figure 1.3a, where it has precise starting and end points, although there is much to suggest that this rigid approximation is not suitable. In many organisations some kind of planning, for alternatives, is ongoing most of the time. One should treat the planning process as a continuous activity, as in Figure 1.3b, and all interested parties should participate in the activity.

The past decades have seen a paradigm shift in our understanding of work and the conditions of work. The perception of people and job design with people in mind is much more holistic, as opposed to pragmatic. A pragmatic approach would tend to aim for a solution regardless of how people are affected by that solution. Nowadays, participation from employees is an obvious necessity. This has a large impact on the job content. This is discussed in this new edition of the handbook. Today, control room operators have larger and more complex jobs. But in most control room situations, operators have no opportunity to envision new perspectives of their industry, and this is a problem for operator and process development. Concerns for environmental and energy husbandry have become more prevalent.

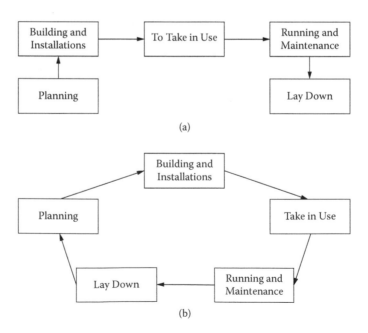

FIGURE 1.3 The ongoing process for planning and building new industries.

1.2   PARADIGM SHIFT: FROM CONTROL
ROOMS TO CONTROL CENTRES

In his classic book *The Structure of Scientific Revolutions*, published in 1970, Thomas Kuhn described the natural sciences from the perspective of models of thinking where one paradigm would be the universally accepted view. Paradigms encourage people to see things in a certain way. This happens in industry and business, as well as in academic disciplines. Thought and action continue along accepted pathways. Any alternatives to the accepted way of seeing and doing are regarded as irregular, unacceptable, or downright wrong. That is, until someone or something changes the prevailing worldview. Technologies often disrupt existing paradigms. Information technologies (IT) such as the Internet, the widespread use of personal computers, and the phenomenal growth of e-mail, search engines, and information-sharing facilities (such as YouTube, MySpace, and Facebook) disrupt many existing paradigms as people adapt to new channels of information. Today we face a new paradigm shift. The control *room* concept seems to be on its way out. In new industries and in professions where new technologies are used, control *centres* have become the norm. We will therefore talk about control centres as these incorporate a wider scope of supervision, control, and development. In these contexts we will see greater complexity. For example, communicating and control systems will have complex hierarchical processes. Whether local, regional, national, or global, the processes will enable communication to take place in real time between these different levels of the hierarchy. Operating between these hierarchies will be networks of subsystems. A classic example is the reservations systems of airlines. These were among the first data and information systems to be globally networked, allowing user access from anywhere within the network.

The 'control room concept' is spreading to many new areas. For example, the food processing and food technology industries are now included in the traditional control room concept. There are many obvious advantages to making these industries automated and limiting direct human involvement that leads to, for example, improvements in levels of hygiene quality. However, a problem of introducing automation in this context relates to the difficulties in measuring different quality-related factors without direct human involvement. Many aspects of food processing are dependent upon tacit knowledge of professionals and, often, such knowledge is the preserve of specific 'expert' individuals. In brewing, winemaking, and cheese fermentation, the brewers, vintners, and cheese makers often rely upon their 'nose' to gauge quality.

In recent years, there have been a number of completely new areas of control rooms—or, rather, control centre—types of applications. One very fast-growing application is in surveillance and security control centres. These incorporate much more than 'policing' of urban, business, and residential districts. Yet another area is that of financial control and trading centres. These began to emerge to some extent more than ten years ago. Today most large companies have control centres to improve the quality of financial planning and to provide real-time data on market movements. Companies such as Reuters and Bloomberg have successfully exploited the need of the financial services industries for such data.

In the public management sector, this past decade has seen the emergence of e-government systems. However, while development has been rapid, these systems have tended to be in isolation. Today more and more of the work of government departments is coordinated in e-government systems. Often e-government systems are very clearly separated from the departments working in traditional ways. One risk here is that the two processes will continue to work in parallel and the government and its departments will have duplicate operations. In this scenario, there is a need to create some kind of control centre that clearly will be controlled by the top management of the organisation. In the near future, one more-than-likely scenario is that national systems will interconnect all (or at least most) of the government services within a country. In this scenario there could be virtual connectivity between the member systems at different levels. Different national systems could be interconnected to create clusters of regional systems. Existing systems of e-government and multinational systems such as the different agencies of the UN and the EU are likely to become embryonic control centres on a global scale. A key issue for government and nongovernmental organisations (NGOs) is the need for high-quality information that is formally correct. In this, a critical issue in the future will be to decide how to handle the different legislations (for example, regarding openness and public availability of government information) that are involved when control systems have a global reach.

REFERENCES AND FURTHER READING

Amerine, M., Pangborn, R., and Roessler, E. (1965). *Principles of Sensory Evaluation of Food*. New York: Academic Press.

Åstrand, O., and Rodahl, K. (1977). *Textbook of Work Physiology*. New York: McGraw-Hill Books.

Berglund, L. (1982). *Systemframtagning av högautomatiserade styroch reglersystem med hänsyn till driftsäkerhet och livstidskostnad*. EKZ 7, Stockholm: Statens Vattenfallsverk.

Berns, T. (ed.). (1984). *Ergonomics Design for Office Automation*. Stockholm: Ericsson Information Systems.

Helander, Martin. (2005). *A Guide to Human Factors and Ergonomics* (2nd ed.). Boca Raton, FL: CRC Press.

Johansson, Gun. (1982). *Social Psychological and Neuroendocrine Stress Reactions to Mechanized and Computerized Work*. Paper presented at the 18th Wissenschaftliche Jahrestagung der Deutschen Gesellschaft für Sozialmedizin.

Kroemer, Karl H.E. (1997). *Fitting the Task to the Human: A Textbook of Occupational Ergonomics* (5th ed.). Boca Raton, FL: CRC Press.

Kuhn, Thomas. (1970). *The Structure of Scientific Revolutions*. Chicago, IL: University of Chicago Press.

Leamon, Tom B. (1995). The Evolution of Ergonomics. *Risk Management* 42, 9 (September), 47–52.

Rowan, Marilyn, P., and Wright, Philip, C. (1994) Ergonomics Is Good for Business. *Work Study* 43, 8, 7–12.

Shackel, Brian. (1997). Human-Computer Interaction: Whence and Whither? *Journal of the American Society for Information Science* 48, 11, 970–987.

Sorge, A., Hartmann, G., Warner, M., and Nicholas, I. (1982). *Microelectronics and Manpower in Manufacturing*. Berlin: International Institute of Management.

Weiner, N. (1961). *Cybernetics, or Control and Communication in the Animal and the Machine.* New York: MTT Press.

Welford, A.T. (1964). *Ergonomics of Automation.* London: HMSO.

Wilson, John R., and Corlett, N. (eds.). (2005). *Evaluation of Human Work* (3rd ed.). Boca Raton, FL: CRC Press.

2 Models in Process Control

Toni Ivergård and Brian Hunt

CONTENTS

The literature describes relatively few models that have been developed specifically for control of process industries. The majority of the existing models are based on controlling aeroplanes, ships, and the like. Various types of human/machine models and operator models have also been discussed in relation to the control of nuclear power stations.

Some general models for human/machine control are described first. There then follows a presentation of some general thinking behind dynamic models for integrated human/machine control systems (Figure 2.1A). Here the aim is to develop the technical control theory through to a more detailed discussion of models of the operator's cognitive processes in control systems and special control room applications (Figure 2.1C), together with the operator's mental experience-based model of the technical process (Figure 2.1B). Finally, some comparative commentary on the value of the different types of models is presented. Technical models of the actual process are not covered here but can be found in Lowe and Hidden (1973). More recently, many authors have presented new models of relevance for process control. Typical examples are the work of Schön (1982) and Raelin (1997); this aspect is discussed in Chapter 11.

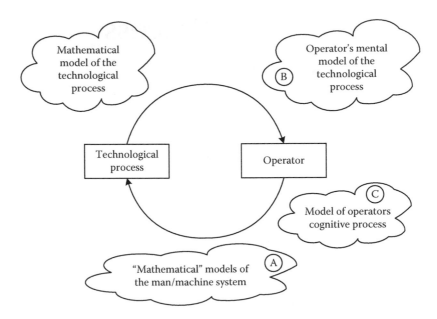

FIGURE 2.1　General model for human/machine control.

2.1　GENERAL MODELS

Before entering discussions about the different forms of models for human/computer systems in the process industry, it is important first to look more closely at the reasons behind the demand for computerisation and automation.

2.1.1　Motives for Automation and Computerisation

As in other industries, there is usually a requirement that investments must be economically profitable. One expectation of computerisation may be that it will lead to reductions in personnel (and thereby reductions in cost) or that the demand for qualifications among the staff could be reduced, thereby also reducing long-term costs. There may also be expectations that computerisation will increase income by, for example, increasing production capacity or bringing improvements in product quality. Another important reason for computerisation may be to reduce costs by reducing the disturbances in production and thereby achieving more efficient use of production capacity.

Apart from these economic profitability requirements, there are other motives for investing in computerisation. Pressure to automate may be due to the demands from employees and/or authorities in order to reduce the risks to workers or the community in situations where there is a danger of exposure to substances that could damage a person's health or the environment, or to eliminate the risk of explosion.

Even if the economic profitability requirements are high within a process industry, it is no longer common to have great expectations that computerisation will lead to great reductions in staff numbers. Studies carried out by Ergolab* have not shown any direct relationship between the degree of computerisation and staffing levels (Ivergård et al., 1980). On the other hand, considerable increases in production volume have often resulted from increasing the degree of computerisation. This is typical, for example, within the paper industry.

In the electricity generating industry a relationship has been suggested between levels of automation and staffing. This is shown in Figure 1.2. When automation starts, drastic reductions can be made in the number of personnel. As automation continues, only smaller personnel reductions can be made, and eventually the curve tends to level out. One can even speculate that, if the automation level is pushed very high, the personnel requirements may actually increase again. By automation level in this context we mean the size of the investment, which is made for controlling the process using different types of automatic aid (for example, computer-based control systems). Even within process industries today, such a high level of automation has been reached using traditional technology that no further radical staff reductions could realistically be expected.

Within the electrical power industry, and to some extent the petrochemical industry, the most important reason for automation is thought to be that of achieving safer and more reliable production. In some of the petrochemical industries studied (Ivergård, Istance, and Gunther, 1980), the demand for production quality was an important motive. Lately, energy savings and environmental concerns have become important motives for automation.

The reason that there is no great reduction in staffing levels often lies in the fact that a certain minimum manning level is required to handle any disruptions or emergencies that may occur. In addition, the requirement for maintenance, service, and monitoring in a plant does not diminish as a direct result of automation. It is even quite likely that these tasks will tend to increase. Although computerisation may not substantially alter the size of the workforce, it inevitably brings about major changes in work roles. The traditional control room operator's job changes extensively, and maintenance staff and their counterparts also discover that their job content alters. It is important to determine which tasks can suitably be combined into the new working roles required when introducing computers.

Studies carried out during the 1970s and 1980s show that there is a tendency for people to retain their traditional work roles, with special operators to monitor the process and other operators needed for maintenance, service, and repair. Planning and optimisation tasks connected with the control system become the work of technicians and management staff. This form of role demarcation in conjunction with computerisation has several disadvantages, particularly for the control room operators. It often results in the operator doing a daily job requiring lower levels of qualification, but it does not lead to the recruitment of operators with fewer qualifications. This is because occasionally—for example, during interruptions or under special operational conditions—the operators need at least their original level of competence to deal with these circumstances. Because people do not need to use the abilities that they are required to have, they get too few opportunities to practice

---

* An international research and consultancy company active in the field of ergonomics.

their job skills to be able to cope with the difficult peaks in the work. In addition, the level of attention becomes much too low to allow for a suitable degree of readiness to cope with demanding situations when they do arise.

Some of the companies studied (Ivergård, Istance, and Gunther, 1980) attempted to create new job roles consisting of combinations of tasks in process plants. This has been shown to have many advantages, including increasing competence. In most of the plants studied, it should also have been possible upon computerisation to require control room operators to perform more of the production planning and optimisation tasks, although examples of this are not forthcoming.

During the past decade there has been a clear tendency to increase the competence levels among control room operators. In studies carried out in the paper and pulp industry and in the production of hydroelectric power, most control room operators have advanced university technical training. There are also discussions for the 24-hour shift teams to include computer experts. There are still different opinions if the operators should mainly be working in the control room or whether their work tasks should also be carried out in the plant. In some segments of the paper and pulp industry, there seems to be a trend for operators to also work outside the control room,

It might be imagined that with computerisation, companies would, at least as a secondary aim, attempt to create interesting and meaningful jobs for the operators. However, there seem to be no examples of companies using the introduction of computerisation to attempt to enhance the quality of work. It is not possible to elaborate this issue here; we shall only deal with questions which are of more direct relevance to production.

It is usually suggested that it is desirable for operators to have some control over their own work and that the work itself should not be too 'machine-bound'. It is also desirable for the work to be experienced as a whole, and for the operators to feel that they are performing a fulfilling job.

A characteristic of process control work is that one is working with a continuous operation, which means that the operators can seldom experience the type of satisfaction which comes from performing a complete task. On the other hand, a similar form of satisfaction can be obtained from knowing that the required qualitative and quantitative standards have been achieved in the work. One company studied gave the operators regular and clear information on the extent to which the qualitative and quantitative requirements were being fulfilled in order to provide work satisfaction (Ivergård, Istance, and Gunther, 1980).

The degree to which the operator is bound to the machine is largely determined by the design of the process. A continuous process—as distinct from a batch process—necessarily means that it is the process itself which determines the time and pattern of the work, and the operator must work according to the requirements of that process. The time scales are dependent on the dynamic time characteristics and response times of the process. In a process industry with continuous operations there is thus a high degree of time-bound work. This form of time-bound control is seldom experienced by the operators as being particularly stressing, especially where the processes have a high inertia, which means in practice that it is never second-by-second control, or even minute-by-minute control. What is more important is the degree of instructional control (that is, having to resort to reading instructions in case of disruptions or similar events). Here, different types of process industries show considerable differences. The operators in industries with 'high' demands for safety often have a high degree of instructional control as a form of safety net for the planners and management. This is particularly true in the nuclear and the electricity industries. Other types of process industries, such as the petrochemical industry, have a very low degree of instructional control by virtue of the fact that the operators are largely allowed to determine for themselves the actions that should be taken under different circumstances.

2.1.2 EXAMPLES OF COMPUTERISATION IN PROCESS INDUSTRIES

It is not difficult to design a model that reflects in a general way how human/machine interaction is usually affected by computerisation as normally carried out in process industries. Moreover, at a general level, the model would not be particularly complicated. The usual reason for working with such a grossly simplified view of human/machine interaction is that the goals are too limited at the planning stage—being directed mainly towards economic profitability with no deeper understanding of the human role and its contribution to efficiency in such systems. Successively, we have a better understanding about a more meaningful and effective use of the human operator in control rooms. One can see a clear change in focus from operator as a monitor of the process to operators working to optimise the production while at the same time reducing the risk of errors and accidents. Furthermore, the operator has new tasks to redesign and develop the control system and the production process.

The model shown in Figure 2.2 is often used as a basis for discussion of human/machine systems. This shows, in a simplified way, how the human operator interacts with a process via a display (information device) and a control device. When computerisation takes place, the human operator is replaced by a computerised control system according to Figure 2.2. Figure 2.3 illustrates the principle that the human operator's primary role becomes that of supervising the process using the computer

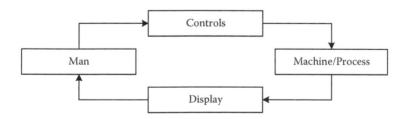

FIGURE 2.2 Simplified model of the human operator's role in control systems.

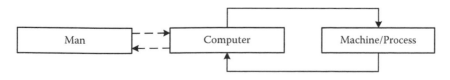

FIGURE 2.3 Simplified model of the human operator's role in a computerised control system.

system, and in certain cases (usually when something goes wrong with the automatic control) to intervene and control the system in the traditional way. The dashed line in Figure 2.3 represents the indirect control of the process via the computer; the solid line represents the interaction between the computer control system and the process as it occurs normally.

As a method, Figure 2.3 is a gross simplification. No computer system has yet been installed that completely takes over the majority of the control functions in a process. A process industry consists of a large number of smaller subprocesses, and different subprocesses may have several input and output variables that can sometimes be regulated separately from each other. The various subprocesses, or parts of them, may be automated successively by installing analogue control systems. Initially, these systems of regulators could be sited and controlled locally. As automation proceeds further, the regulators may then be brought together into a common control room for the whole process (remote control). In this way, automation progresses as more and more regulators are brought into service, as shown in Figure 2.4. The operator's task then becomes one of setting up the set values on the regulators and monitoring the processes that are being controlled automatically. In addition, of course, the nonautomated functions still need to be controlled.

The first stage in computerisation is to start to change the analogue control system to electronic microcomputers. The Honeywell TDC2000 was a first, very early, and typical example where the analogue regulators are replaced by a large number of microcomputers that can then be connected together into a common system (using visual display units [VDUs]) for monitoring and for insertion of the set values via a keyboard. The TDC2000 system has been continuously developed and updated and is still in use. A large number of similar systems are today available for use in, for example, power production, the paper and pulp industry, network supervision, and so forth. A similar system has been developed for information management control (for example, knowledge management).

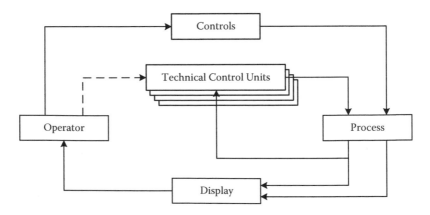

FIGURE 2.4 As automation progresses, the human task becomes one of checking standards and monitoring the automatically controlled process.

Sometimes, the microcomputer makes possible more advanced control and coordination of the various control loops, for example, sequence regulation and control according to certain programmes. Figure 2.5 shows the case where a number of control functions have been transferred to computer.

This common method of computerisation often resulted in reducing the worker's active role in the system to such a degree that his or her best characteristics (flexibility, experience, long-term memory, and job skills) are no longer required and the disadvantages of the human as an operator are accentuated (for example, the inability to maintain long-term attention in so-called vigilance situations). Thus the worker's role and position within the system are not suited to his or her intrinsic abilities. A more recent approach includes the operator in the development of the system. We are slowly seeing a paradigm shift from the 'rational man' approach to a more participative way of working.

Human beings could be allowed to remain in the main control circuit but provided with improved information that can be produced by the computer (Figure 2.6). The computers could, for example, be used to take the various readings and to make calculations on the basis of these. One valuable aid may be, for example, to have some form of automatic model of how the process functions. Based on this process model and the different measured values, the computer can calculate (predict) the changes that will occur in the process if there are no further possible control actions

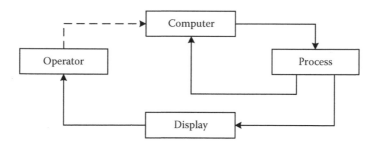

FIGURE 2.5 Computer control.

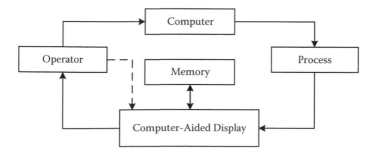

FIGURE 2.6 Computer-aided display.

as 'trials', as shown by the dashed line in Figure 2.5. In this way, one can determine the effects of different control actions on the process. The action that gives the best outcome can thus be found with the aid of the automation and computerisation of information provision.

In a similar way to that shown in Figure 2.7, computers may be used to process and automate the controls used by the workers. In this model the operator is still a part of the control circuit. Traditionally, the steering of a ship is performed by the helmsman, who estimates the angle of the rudder. When steering very large ships, for example, it is very difficult to judge what a particular rudder angle will produce in terms of turning radius of the ship. Based on the desired course, the helmsman sees on the chart or the radar screen not the rudder angle, which should be assumed, but the radius at which the ship should turn. Using this as background, steering aids are being developed today where, instead of giving orders to the steering machinery to set a particular rudder angle, orders are given to a computer on the radius the ship should turn. Knowing the dynamics of the ship, its speed, shape, and the size of the propellers, and so on, the computer chooses a sequence of suitable rudder movements. The computer may also make use of signals giving the turning rate as a type of feedback in order to calculate more reliably the best rudder movement. Thus, the best possible correspondence between the measured and the desired turning radius may be achieved.

These two simplified models for computerisation (Figures 2.6 and 2.7) have a different character from the first (Figure 2.5), which is the most common at present. There have been very few examples of the latter type of computerisation within the process industry. Examples of applications of these forms of computerisation have to be sought in other areas, such as in the control of ships and aircraft, where they can be considerably better in certain connections. This leaves the more interesting and meaningful jobs for the workers, while at the same time they remain at the centre of the process, and their job skills are better used.

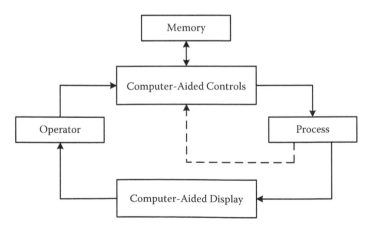

FIGURE 2.7   Feed forward aid.

2.2 BASIC CONTROL CONCEPTS AND HUMAN/MACHINE MODELS

A general model of human/machine systems will be considered (see also van Cott and Kinkade, 1972; Ivergård, 1982).

Control is defined in international standards as *a general concept to denote purposeful influence.* Figure 2.8 shows schematically the main components of a control system. The system has input quantities (y_i) and output quantities (y_o) respectively. The relationship between these input and output quantities is determined by a law or transfer function that is dependent on certain parameters (*ps*). The transfer function of a system may thus be expressed as an equation:

$$y_o = f(y_i, ps) \tag{2.1}$$

A system can be described if the inputs, the transfer function, the parameters, and the output quantities are known. Where one or more of these is unknown, various methods may be used to define them. The following combinations of known and unknown quantities can occur:

1. Inputs, transfer function, and parameters are known and the output is required, for example, in the evaluation of a system under design.
2. Outputs, functions, and parameters are known, and the inputs are required. The method used in this case is known as diagnostic and is used, for example, by doctors trying to find out the type of disease, i.e., trying to determine the reason (input) for the symptoms (output) shown by the patient.
3. Inputs, outputs, and transfer function are known, and the parameters are required. This method is known as identification, and is used, for example, when one wants a mathematical description of a particular event.
4. Inputs and outputs are known, and both the transfer function and the parameters are required. The method for this type of problem is called the 'black box' technique, and is the one commonly used in the description and testing of very complex technical systems such as computers.

In process industry control systems, all quantities are more or less well known, depending on how well 'identified' the system is; in other words, they are like option (3). A skilled operator or a well-developed computer control system 'knows' the various parameters well. In another case, one may be working with some form of 'trial

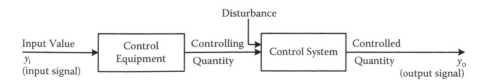

FIGURE 2.8 Diagram of simplified control system.

and error' philosophy. In practice there are always some unknown factors—disturbance quantities (y_n)—for which the operator/control system must compensate.

Two types of systems are concerned in control: the controlling system (control equipment) and the controlled system (process) (see Figure 2.8). The control equipment controls with the help of the controlling quantity, so that the desired value (set value) of the process can be maintained. The true value which the process returns is called the actual value. The process is also affected from outside, and this effect is known as the disturbance quantity.

2.2.1 Open and Closed Controls

In open control, the control equipment is not affected by the object or process. Certain types of control equipment in central heating systems are examples of open control (see Figure 2.9). The control equipment has the job of reducing the temperature in the water used for heating at night, according to a preset programme. The measuring device, for example, senses by means of electrical contacts switched on or off by pegs from the control device, the stage in the programme that has been reached. The contacts 'inform' the central link, which processes the information. After processing, certain of the input conditions given by the measuring device will result in a command to the control device (for example, to start valve motor 1). Note that no signal from the process/control object affects the control equipment.

It is very difficult using open control to maintain the actual value at the set value. This can only happen when the object is not exposed to any external disturbances, or where these are well known and can be taken into account in advance. Where these conditions are not met, large deviations from the desired values must be allowable in order for open control to be used.

Closed control (see Figure 2.10) has a system where the output signal (actual value) is fed back to the input signal (desired value) and compared with it. The input signal is, in certain cases, the desired output signal. When this required actual value, the desired value, is compared with the true value fed back, a difference results if the actual value deviates from the set (desired) value. After amplification via the control device, this deviation then goes in as the control quantity to control the process or object.

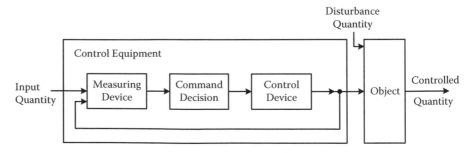

FIGURE 2.9 Open control: Note that there is no feedback from the output of the controlled object to the controller, but only within the controller itself. Changes in the controlled quantity thus cannot affect the control equipment.

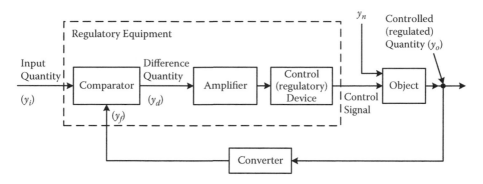

FIGURE 2.10 Closed-loop control, also known as regulation. The regulated quantity affects the controlling signal.

The controlled system is also affected by external disturbances (y_n). The signal (y_o) goes out from the system, is measured, and fed back.

When the comparison between the input (y_i) and the feedback signal is an additive, this is known as positive feedback. This type of feedback, which gives an accelerating change of the output signal, is used, for example, to achieve the fastest possible changes of output signal. When the comparison between y_i and y_o is subtractive, we have negative feedback. This means that y_d (difference quantity) decreases when the output y_o increases, and thus attempts to stop y_o rising. Negative feedback is the most common form of feedback. One well-known example of a control mechanism with negative feedback is the regulation of the water level in a WC cistern by a ball-cock valve. Figure 2.11 shows a diagram of a cistern with its ball-cock valve and the corresponding signal control diagram (a block diagram).

A distinction may be made between closed control systems,* which only have to compensate for outside disturbances and not for changing inputs, and closed control systems which are not exposed to outside influences but which only attempt to follow changes in input quantity. The first type of regulation system (where the input value is not changed) is known as constant regulation. The second type, with variable inputs, is known as servo-regulating. A servo-regulating system only has the task of trying to achieve an output that follows as closely as possible the varying input level. The WC cistern described above is an example of constant regulation, which tries to keep the output (in this case, the water level) constant regardless of outside disturbances (that is, the water being flushed out of the cistern). Most regulation systems in process industries function as constant regulation systems (where the set value can be changed). At start-up, however, a certain degree of servo-regulation takes place.

---

* According to the Swedish Electrical Standards (SEN 0106), a closed control system is not a regulated system when the control equipment consists of a person. The authors' opinion is that this is not applicable here, where the same terms are used for both human and technical components. The concepts of a regulated system are therefore also used here for manual control with feedback.

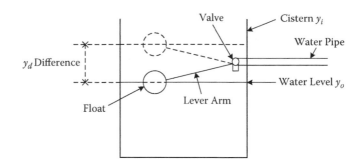

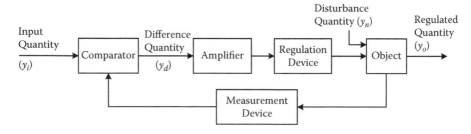

FIGURE 2.11   Regulation of water level using a float valve (constant regulation).

2.2.2   HUMAN/MACHINE CONTROL SYSTEMS

A human operator can become just as important a component in a regulatory or control system as the electronic or mechanical components. Constant regulation (where the operator puts in new set values at certain intervals—running orders) is normally found in industrial processes. The concept of 'controller' should only be used where the worker is an 'online' part of the regulatory system; otherwise, the more general term 'operator' is usually used.

When someone is controlling a car, he or she has to carry out a number of mental differentiations and integrations, depending on the response produced to the input quantity (see Figure 2.12). Similar conditions exist for many other control tasks, for example, steering a ship. A human being's ability to carry out these mental processes successfully is relatively limited. Two French researchers (Tarriere and Wisner, 1963) showed that fatigue produces very strong oscillations when controlling a car. The ability to carry out the necessary integrations and differentiations clearly worsens with fatigue.

The task of controlling a machine can be made easier by introducing various aids (off-loading mechanisms) for performing the integrating and differentiating tasks. There are two main types of such mechanisms:

1. Influence the controlled object directly, and thereby the actual value ('A' mechanisms—Aiding);
2. Influence the feedback, and thereby the information to the human operator ('Q' mechanisms—Quickening).

(a) The vehicle's control response

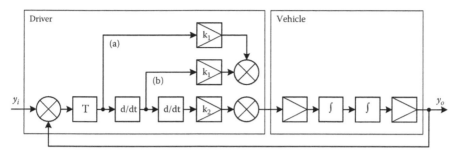

(b) Driver-vehicle control response

T = Time delay (man's reaction time)

FIGURE 2.12 Control response of a vehicle (a) and the driver-vehicle control response (b).

Introduction of an 'A' mechanism, as shown in Figure 2.13, brings about a reduction in the worker's load, as certain calculations no longer need to be carried out. Also, the aiding mechanism, which lightens the load concomitantly makes the system faster. Without this mechanism, the operator's control output of the system is affected. As the quantities within the system are fed back continuously to the operators, they will be immediately aware of changes, which occur as a result of their actions.

Simplification of the controller's task using an 'A'-simplifying mechanism works preliminarily in two ways:

1. By relieving the operator of certain operations
2. By giving information more rapidly about changes in the process

'A'-simplified mechanisms cannot be used where it is impossible to make the technical changes which will directly affect the output quantity of the controlled object. These output quantities may depend on quantities that cannot be controlled, such as wind, waves, and similar aerodynamic and hydrodynamic relationships. In such cases, a 'Q'-simplifying mechanism can be used. 'Q'-simplifying mechanisms do not work directly on the output quantities of a system, but they change and simplify the information which is presented to the controller. There are different types of 'Q' simplification:

1. Predictive 'Q' information
2. Complete 'Q' information
3. Partial 'Q' information

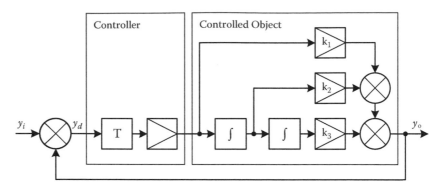

FIGURE 2.13 Introduction of aiding—the 'A' mechanism.

The need for 'Q'-simplified mechanisms occurs in many industrial processes, such as the starting up of a process where certain values (for example, temperatures) have to be increased according to a particular time plan. Such systems often have considerable inertia, which means that there is no direct relationship between control movements and the output of the system, and it takes a long time before any result can be read from the instruments.

The operator has a control level whose movement (input quantity to the process) is integrated a number of times before the desired temperature change (output quantity from the process) is obtained. In such cases the desired temperature sequence cannot be set. In order to improve the performance of the operator, the temperature (output quantity) can be differentiated a suitable number of times, and this information will then form the basis of an automatic calculation (prediction) of the temperatures at various times in the future for different degrees of control movement. These expected temperatures can be presented as a curve on a screen, on which the desired temperature sequence can also be displayed. Figure 2.14 shows the required sequence of values marked as a line on the screen. The prediction of the course of the change for the actual control movement is automatically calculated and should normally be shadowed by the line which shows the desired sequence. In this way, any deviations can be seen very quickly. This is shown as predictive 'Q' information.

In modern industrial processes there is no direct regulation; the operator sets in the desired values at certain times, and the process is automatically controlled within these values (technical constant regulation). Starting and stopping of the process is done with what are known as group starts, where one control is used to

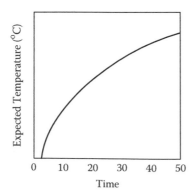

FIGURE 2.14 Predicted values presented on a display.

initiate the start-up, in the correct sequence, of whole groups of motors, valves, and so on. This is often thought to be necessary so that the handling of the process will be sufficiently rapid or accurate. Predictive 'Q' information should probably be used here as an alternative or an addition, as this would provide the operator with a better understanding of how the process works. The work would also be more interesting, which in turn could lead to the process being better controlled.

Figure 2.15 shows a system containing four integrating elements. One example of a machine which operates in this way is a submarine. When a change is made to the depth control of a submarine, it takes a long time before a result is obtained in the form of a new stabilised depth being reached. Also, the movement of the control is not proportional to the movement of the depth control integrated (in this example) four times. Because the movement of the submarine also depends to a large extent on water currents, hydrodynamic forces, and other external factors, the output—that is, the ultimate depth—cannot be influenced by an aiding mechanism which will directly affect the movement of the vessel, and thereby the output quantity. It is also very difficult to control the submarine solely with the aid of a depth gauge. One alternative is to feed information back directly to the operator on what is happening to the submarine at different stages, as shown in Figure 2.16. This is known as complete 'Q' information. Instruments continuously measure the depth, vertical position, change of depth, or diving angle with time (first time derivative of the output), rate of change of diving angle (second derivative of the output), and the acceleration of the diving angle (third derivative of the output).

In order for the 'Q'-simplifying mechanism to function correctly, the constants in Figure 2.16 must be very carefully determined. There is no complete theoretical method for determining these constants, but their optimal values can often be deduced with the aid of analogue computers set up to simulate the system in question. The derivative which is weighted with the output quantity to produce the feedback need not be a time derivative but could also be a position derivative, for example. Where it is not possible to measure the different derivatives directly, they can be calculated indirectly using, for example, electronic differentiation.

As a summary of the human ability to function as a continuous controller in a system, it maybe stated that the tasks performed should be as simple as possible in purely mathematical terms. If possible, they should not be more complex in practice than a person functioning as a simple amplifier, that is, no differentiation or integration is required. With the aiding mechanisms described here, it is feasible to introduce tasks specially matched to a worker's performance abilities.

The types of models described so far represent simple control circuits, for example, control of speed in a car or the position of the car on the road. They have their

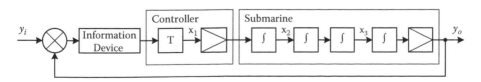

FIGURE 2.15 System containing several integrating elements.

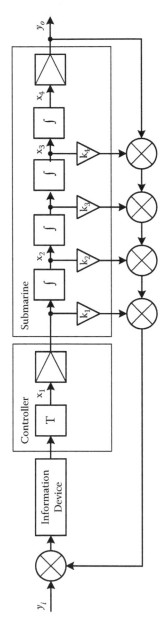

FIGURE 2.16 System feeding partial information back to the operator.

primary practical applications in the control of, for example, aeroplanes, cars, and ships. As far as the control of more complex processes within the process industries is concerned, these types of models are only of importance for the understanding of the control of the individual process parameters. A process industry, on the other hand, consists of a large number of parameters, commonly several hundred. When the operator is functioning directly in the control loop, however, it is important to understand these theoretical basics first.

2.3   MODELS OF COMPLEX HUMAN/MACHINE SYSTEMS

There have been many different attempts to produce a model of a human being as an information processor. Many of these have attempted to describe a person mathematically as an information processor in complex systems (see Timonen, 1980, for an overview of these). It is typical of these mathematical models that they can only handle a small part of human behaviour and as such their practical use is very limited. They may be of some value in describing a particular form of behaviour in very critical and important situations—for example, an instrument monitoring the start-up of an aeroplane—but the value of such models is probably limited to helping in the structuring of the problems. On the other hand, the actual mathematical calculations are of less interest. Singleton (1976), for example, stated that the mathematical descriptions are difficult and provide little extra information for the prediction of the operator's behaviour. They are also often based on simulator studies, which always deviate to some extent from the real-life situation. Thus, the nonmathematical models of a more general nature will be described in this section.

The aim of this type of model is to increase understanding of how a human works as an operator in a control room and to thereby improve and optimise the design and functioning of the control system. Models that describe a human being's information processing capabilities and associated cognitive processes will be considered first. A discussion of the well-known mental models that humans use to assess the actual physical process is also presented (Bainbridge and Beishon, 1964; Crossman, 1965).

2.3.1   Models of Operators as Components in Complex Systems

The following description is primarily limited to the role of the operator as a system component. First it is helpful to examine the operator's ability to fulfil the goals of the system. Of course, many other factors influence the operator and thus the operator's work. The operator as a contributor to, as well as a part of, the complex process control system is only one aspect of the possible influencing factors. As a social being, the operator is influenced by background, leisure, and family circumstances, for example. This more complete view interacts in different ways with the individual's role as a system component (this interaction is discussed in Chapter 8).

The classical way of describing human information-processing abilities often starts from a model such as that shown in Figure 2.17. This type of model is based on descriptions by, for example, Welford (1968) or Ivergård (1981). The operator uses mental processes to convert the various sensory impressions and signals from the environment. These signals are structured in what is known as perceptual

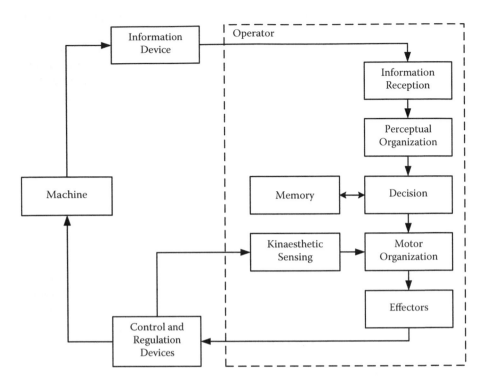

FIGURE 2.17 Human-machine system with a highly-simplified description of the human operator where, for example, the various forms of feedback mechanisms have been omitted.

organisation whereby the primary signals are converted into some form of meaningful whole, for example, letters, pointers, or what they indicate. Based on these structured signals, the operator makes different types of decisions. The decisions are taken on the basis of various forms of stored 'set values' (memory) of what the operator feels the signals should actually be in each particular case. The decisions have an effect on the 'motor organisation' in the cortex. This forms a ready-made and stored programme in the human brain capable of carrying out various types of motor activities under different circumstances. The motor programmes then activate groups of muscles in predetermined sequences.

Using the above model as a starting point, it is possible to differentiate diverse types of skills. The ability to control some parameters directly 'online'—for example, to hold a car on a winding road—is called *sensorimotor skill*. A typical example of a *perceptual skill* is the ability to understand a three-dimensional world when represented on maps or radar screens. Perceptual skills also include many forms of inspection work in industry, for example, inspection of electronic circuit boards, as well as engineers listening for faults in a diesel engine. Many types of perceptual skills involve what is known as vigilance work situations. This means that the signals to be detected, such as faults in the process, are very few, unclear, and occur randomly. *Vigilance skills* are thus a special form of perceptual skill.

It is typical for all types of skills that the ability to carry out a task correctly is developed through training. However, in order for the skill to develop, training must often be of a practical nature. It is not sufficient just to know how to do something, but one must also be able to do it in practice. It is possible using written instructions to explain, for example, how a radar picture should be interpreted. On the other hand, the ability to read a radar picture correctly is only developed after many years of work actually reading radar pictures. Another typical example of a perceptual skill is that of lookout on a ship. Practically always, the older seaman will see other ships earlier than the inexperienced one. This may be partly explainable by differences in visual ability, but is probably primarily attributable to a better developed perceptual skill.

When information processing is mainly concerned with decision making—that is, cognitive tasks—this is referred to as *cognitive skill*. Here, the degree of difficulty of perceiving the incoming signal may be relatively simple. The difficulty lies in manipulating the incoming signals suitably, putting these in the correct context, and drawing the correct conclusions from them. This becomes a question of interaction partly between different parts of the long-term memory, and partly between long-term and short-term memory. This is the case, for example, where an operator in a control room has to interpret correctly a particular reading on an instrument. A change in a reading may depend on many parameters. A skilled operator has a picture of how the process works stored in his or her long-term memory. Faced with incoming signals about changes in the process, the operator continuously updates the mental picture. In any situation, there may be many sensory inputs that contribute to this updating. For example, there may be signals from different instruments in the control room, or the operator may actually have noticed at firsthand the various changes that occur in the process (such as increased wear, beginnings of leaks, and so forth). Skilled operators build up a detailed and accurate mental picture (process model), which allows them to understand clearly the meaning of the different changes shown on the instruments and to make appropriate decisions accordingly. Planners in different fields work in a similar way. A subtle difference is that, instead of having a model of an existing reality which is continuously being updated, they have a model of a conceived, planned reality which may become modified in different ways. The work of planners is often very abstract. However, as human beings have a very limited ability to work solely on an abstract plane, so planners often get help by relating the newly planned system to some known and already existing system. This limitation in the human ability to work with abstractions is a barrier to the creativity needed to find new and unprejudiced solutions to problems. As human beings we find it much easier to extend ideas from existing concepts than to invent new concepts.

In classical manual crafts, such as those of blacksmiths, shoemakers, welders, glassblowers, and so on, *motor skills* are of the greatest importance. It is typical of a well-developed motor skill to have a very well-developed programme of different patterns of movement stored in the motor organisation mentioned previously. In high levels of motor skill, it is usual not to have to depend on seeing the effect of what one does directly but to have some form of stored feedback mechanism, which goes straight back to the motor organisation centre and corrects and adjusts the movement pattern to suit the actual circumstances. A typical example of a motor skill is that of typing on a keyboard. At a low level of skill, a person looks for each key and then presses it. The motor organisation in this case is fairly simple. The typist has a mental model that directs the fingers to the correct keys. At a very low grade of skill, the eyes are used to steer the finger onto the correct key. As motor skill increases, less visual support is required in order to find the keys. At a highly-developed level of motor skill, a person develops a ready-made mental model to press not just individual keys but whole groups of keys. Relatively independent of the bodily position of the operator, automatic matching of the muscles occurs in order that they will move correctly and hit the correct key or groups of keys. It is often said that skilled practitioners can do their skill 'blindfolded'. Highly-developed feedback mechanisms are found, for example, in welders or in potters where mental models are converted into physical movements.

Development of a high grade of skill is a necessity if a craftsman is to work rapidly. If there is total dependence on perception and the perceptual organisation and its decision-making process, it would take the craftsman longer to complete the task, and the movement patterns would lack precision and efficiency. A human being is commonly regarded as a one-channel system, and since decision making is a sequential mechanism and only one decision at a time can be handled, this creates a bottleneck. However, the motor organisation with its attendant feedback mechanisms can handle several simultaneous operations in parallel—despite a human being only being able to make one decision at a time.

Rasmussen (1980) suggested that there should be a similar mechanism for parallel information processing on a human's perceptual side before the decision mechanism is engaged. Rasmussen developed a very useful model where he describes the control room operator as consisting of a conscious part together with an unconscious process with particular emphasis on the importance of the unconscious dynamic world model. The model is shown in Figure 2.18. Rasmussen considers (in common with Crossman, 1965), that operators have a dynamic mental model of the actual process stored in their long-term memory.* This mental model is updated unconsciously by external signals and internal changes, that is, by the actual process. When monitoring the process, some form of synchronisation continues throughout; that is, the operator uses an unconscious plan to check whether the mental model agrees with how his or her perception interprets the actual process. If there is any form of error in this synchronisation—that is, that the interpretation of the real-life model does not coincide with the mental model—a signal will be sent to the conscious part of the brain. Based on this model, which has certain superficial similarities to the traditional one of the human as an information processor (see Figure 2.19), Rasmussen and Lind (1982) developed different categories of human behaviour or performance level (sometimes using the word 'behaviour' and sometimes 'level of performance'):

1. Skill-based behaviour
2. Rule-based behaviour
3. Knowledge-based behaviour

---

* This is clearly related to the concept of tacit knowledge (see Schön, 1982).

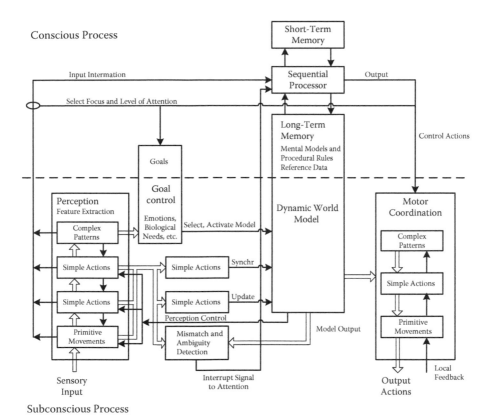

FIGURE 2.18 Schematic map of the human processing mechanism. (Rasmussen, 1980. With permission.)

There is no complete congruence between these different types of 'behaviour' and the different forms of skills presented earlier. As a rough guide, what Rasmussen calls skills-based behaviour corresponds to motor skill and sensorimotor skill; knowledge-based behaviour corresponds approximately to cognitive skill; there is no direct equivalent to rule-based behaviour. On the other hand, Rasmussen discusses a form of skill that is efficiently used to update and synchronise the mental model of the real-life process. This function is probably dependent on a highly developed form of perceptual skill, perhaps in combination with some form of cognitive skill. It is difficult to see the psychological/physiological basis for Rasmussen's division of human behaviour into different groups. Existing bases for division of behaviour in different types of skills are probably a better basis for classifying behaviour.

It is common in connection with job training to talk in terms of factual training (which leads to knowledge) and skills training (which leads to skills/abilities). To a certain extent, these two types of training could be seen as stages in the development of skill, where one first obtains knowledge of the particular situation—for example, the rules and norms (or knowledge training)—and then engages in some form of

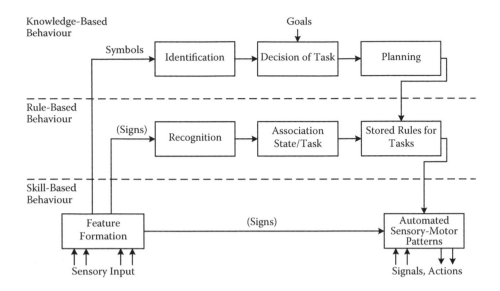

FIGURE 2.19   Different levels of operator performance. (Rasmussen, 1979. With permission.)

practice in readiness to use these skills. There is agreement to some extent between these two bases for division and Rasmussen's division into skill-based behaviour and rule-based behaviour. Skills-based behaviour therefore corresponds to a form of behaviour that is achieved after a longer period of training including readiness training, while rule-based behaviour is achieved after only knowledge training. A wider discussion of these different forms of training can be found in, for example, Ivergård (1982).

Rasmussen (1979a) used the division of behaviour into different groups to some extent as a basis for building up a decision ladder (see Figure 2.20, which is a schematic plan of the sequences of different types of processing from initiation to manual action). This decision ladder has been found to be a valuable aid in structuring and describing control room tasks. Rasmussen and his colleagues have used the method, for example, in collecting data with the aid of verbal questionnaires.

In a project on Computerisation in the Process Industry carried out by Ergolab (Ivergård, Istance, and Gunther, 1980), an attempt was made to use this form of decision ladder, or information-processing ladder, retrospectively in order to describe the various tasks and their parts that were observed. In itself the decision ladder proved to be a useful tool, but it was not possible to use this to generalise between different types of control room work. The ladder was similarly of little use for describing the task that takes the longest time for the control room operator—namely, monitoring the process and updating his or her own mental model of the physical reality. These cognitive processes are in many ways the most important part of the job, as they enable the operator to prepare and create the basis for coping with disruptions in the process. Neither method describes 'skills-based shortcuts'.

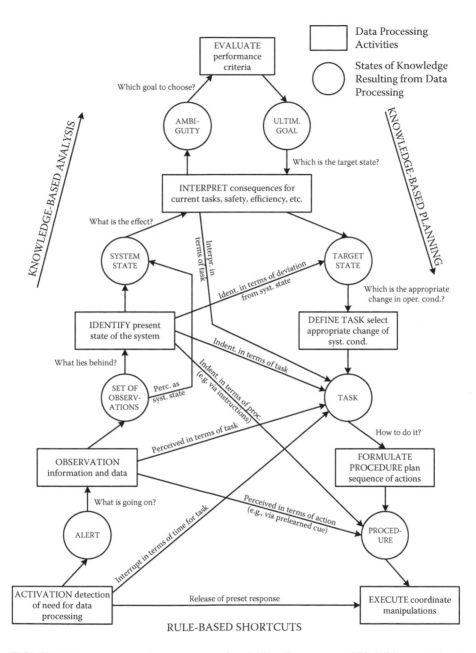

FIGURE 2.20 Sequence of operator mental activities. (Rasmussen, 1980. With permission.)

2.3.2   Mental Models of the Physical Process

As described earlier, the operator has stored in his or her memory a mental model of the physical surroundings. The process control room operator, therefore, has stored in his or her brain's long-term memory a more or less effective and useful model of the process under supervision. The operator continuously updates this mental model by interaction between sensory inputs and the short-term memory. This process probably takes place at a subconscious/tacit level. It is also probable that things are brought to the operator's attention by a perceived lack of agreement between the updated model and the actual state of the process (that is, when the operator perceives that 'something is not quite right'). When this occurs, conscious processing takes over and the operator starts to observe the process in order to analyse this lack of synchronisation. As human beings have a limited ability to process large amounts of information simultaneously, the operator is dependent upon having some form of summary description of the process. The various possible methods for storing this mental model will be examined. The way in which the model of the physical reality is actually stored in the brain is not known for certain. Based on the various methods of describing the process, however, we can make suppositions on the most suitable way to structure the process in the long-term memory. Through having a better idea of the way and form in which the mental mode of the physical process is stored, it should be easier to specify the ways in which the various forms of display devices should present the true status of the process to the operator.

Two main types of model may be used for the graphic description of a process: some type of physical presentation may be made, such as a component flow diagram, or a functional presentation of the process can be given. Singleton (1966) used this basis for distinction. Ivergård (1972) developed the method further in order to describe a process using different functional flow diagrams. Using this method, the starting point is a general system goal. From this goal, different subgoals may be produced where each subgoal consists of a function. The goals may be broken down to different degrees, thereby obtaining functions at different levels of detail. However, these functions must not be broken down in too much detail, as this loses the degree of abstraction as well as the ability to see the functions as a whole. Also, working with too many functions may make the model unusable. If the functions become too generalised there will be too many physical functions and human tasks. When this happens, the model will become of less practical use as, for example, in fault analysis.

Rasmussen (1979a) produced an excellent review of the many different types of conceivable models of how an operator could store structures of the physical process being supervised. Rasmussen starts with a taxonomy of model descriptions and then describes the following models:

1. Model of physical shape
2. Model of physical function
3. Model of functional structure
4. Model of abstract function
5. Model of functional meaning and objectives

The first two are examples of what Ivergård (1972) refers to as physical models. The other three are examples of functionally-orientated models. The model based on functional structure is, in Ivergård's view, an intermediate step between physical and functional description. The logical functional flow diagram described by Rasmussen under the heading of model of meaning and objectives is that closest to the function flow diagram described by Ivergård (1972).

Experience from Ergolab's studies (Ivergård, Istance, and Gunther, 1980) on Computerisation in the Process Industries, however, showed that there is probably a very large variation in the way in which operators build up their mental model of the process. The more theoretically orientated operators, with experience of handling abstract information, have a tendency to form functionally orientated mental models. Examples of such operators are those who consider that they work most efficiently if they just stay in the control room and do not consider that there is any advantage in working in the plant or in being a supervisory engineer (that is, the sort of operator who at certain times goes out in the plant to control, monitor, or even to carry out certain types of maintenance work). Theoretically orientated operators feel instead that they can build up their understanding of the process by being in the control room for as long as possible. On the other hand, there are operators—the majority in Ergolab's studies—who state that they are wholly dependent on being out and doing practical work in the plant and it is in this way they build up their knowledge. This latter form of operator is probably completely dependent on a physically orientated model.

If it is correct that different operators use different forms of models—that is, functionally orientated or physically orientated—they must also use different degrees of synthesis and different degrees of breakdown of the various subprocesses. It is very difficult to use this type of knowledge in order to design display devices for operators.

2.4   RECOMMENDATIONS FOR PRINCIPLES IN THE DESIGN OF CONTROL CENTRES

A general philosophy for the design of control centres and the involvement of ergonomic design factors will now be considered. This section presents a discussion of the principles as a complement to the recommendations for the different types of 'knobs and dials' to be found in some other parts of this handbook.

Principles in the design of control centres can be discussed from three different perspectives:

1. The system design
2. Participative design and action research
3. Usage of handbook data

2.4.1   SYSTEM DESIGN

Obviously, control room operators should have a good knowledge of the process they are operating. They must also be well aware of the task they are expected to carry out. This is also the basis for the planning of the control room and its organisation.

It is assumed that a realistic human resource (HR) deployment and organisation plan has been produced on the basis of the job descriptions. But before deciding the tasks, and therefore the HR deployment and organisation plan, it is necessary to have made reasoned allocations of the functions that are to be automated and those to be performed by the operator. A systematic allocation of functions is a very useful tool to stimulate creative thinking. By using this process, new alternative designs can be innovated. This reasoning ties in with Singleton's (1966) general systems design model. However, it is questionable whether in an overall plan it is possible to generalise about staffing levels and organisation within the process industries and, subsequently, to make design recommendations for the control room hardware (for example, display equipment, controls, working surfaces, workplace design). It is therefore essential to use this form of systematic methods of design as a starting point. It is equally important to include a more participative process of design. (This is discussed more fully below.)

During the course of the project on Computerisation in the Process Industries mentioned earlier, a number of parameters were determined. Determining parameters made it possible to describe different processes in a relatively unified manner and to thereby find a relationship for the staffing and organisation requirements for the process.

The following variables may be used to describe the process:

1. Organisation within the company and in process operation
2. Complexity and degree of continuity
3. Quantitative aspects
4. Fault handling
5. Dynamic aspects
6. Job aids
7. Converging/diverging processes
8. Information and control devices
9. Environmental and layout aspects

Organisation (1) in the process industries is usually orientated towards (a) optimisation of operation, or (b) maintenance and reliability. In the first case (a), the operators' work in the control room is mainly theoretical. In the second case (b), there is usually a combination of different types of work. Here, operators work partly in the control room and partly on supervision and maintenance tasks in the plant itself.

Complexity and degree of continuity (2) imply variables whereby one can describe how difficult it is for the operators to understand and get an overview of the process, and how well integrated are the different parts of the process. In turn, these variables depend on the size of the process and the understandability of parts of the process or the process as a whole. That a process has a high degree of continuity is directly connected with the 'difficulty of understanding' of the process. A high degree of continuity implies a low degree of discrete functional parts (batches) in the process. In other words, as the process is a natural whole, the operator cannot divide it into any natural functional components. A high degree of continuity also gives less time and opportunity for the operator to consider and make corrections compared with a batch process. This is particularly true where the throughput is relatively fast or where it is dependent on several complex process stages.

Quantitative aspects (3) cover the amount of information processing/information supervision that the operator is expected to carry out. In most processes, this is probably a particularly critical parameter in relation to staffing levels. It includes, for example, how many variables are monitored from the control room, how many automatic control circuits are monitored, and how many automatic control circuits the operators handle manually. It also covers the number of alarms and the level of direct involvement in the process. As a rough rule-of-thumb, it may be said that there must be at least one operator per 100 control circuits.

The whole *fault-handling system* (4) is critical for the operators, both in terms of the number of operators and the types of operators needed in relation to job skills, training, and so forth. One also has to be aware of the fact that fault handling should not only deal with severe incidents and accidents. All errors, minor faults, and critical incidents are important warning signals and should be followed up and evaluated as a natural part of the daily routines of process operations.

Dynamic aspects (5) of the system refer primarily to the kind of time-related response the system has to different actions taken by the operators, both manually and in the form of defining set values on regulators. Response times of control/computer systems are also of interest. Systems with high inertia and long response times make the process more difficult to control (see also the introduction to Chapter 3).

Another important factor in determining the level of staffing and type of personnel needed is the type of *job aids* (6) available. This factor also relates directly to hardware design. To a certain extent, this factor is also directly dependent on the choice of automation level (type of allocation). For example, both the staffing situation and the character of the job itself will be affected if operators rely on handwritten log books or on computer memories with printers. The use of predictors and simulators is also of considerable importance. However, this form of job aid has been found to be very uncommon in process industries. Simulators in the control room (on-line or off-line) increase the possibility for operators to build up their control skill and also for optimising the running of the system. The last three variables have not been dealt with here as they are dealt with in other parts of this handbook.

2.4.2 PARTICIPATIVE DESIGN AND ACTION RESEARCH

Action research (AR) is a methodology that allows employees in an organisation to enquire into problems that they perceive in their workplace in order to resolve these problems 'from bottom up'. Action research is not new. The basic methodologies were proposed by Kurt Lewin (1890–1947), who is generally regarded as 'the founder of modern social psychology'. On Lewin's death, later scholars developed his conceptual theories into what became know as the 'classical' model of action research. The Tavistock Institute in London is especially credited with developing Lewin's theories in work conducted there in the late 1940s and in the 1950s.

A key concept in AR is the belief that people accumulate knowledge from their everyday experiences. A second key concept is that people develop mental processes for organising their accumulated knowledge and are able to use the knowledge to resolve problems that they encounter in their daily lives (see discussions in Greenwood and Levin, 1998; Dickens and Watkins, 1999; McNiff and Whitehead, 2000). Employees who use AR in their workplace examine their own work practices and investigate how they might apply these practices to improving their work tasks. An example would be operators who monitor their own work processes and use the findings to improve the existing way in which they do their own work tasks. An added benefit would be if the operator shares these improvements with his or her colleagues. Another key concept for conducting AR is that employees work in an environment that allows participatory engagement with their work processes. By definition, AR requires a workplace environment of employee empowerment coupled with high levels of tolerance for error. (A relevant discussion can be found in Dickens and Watkins, 1999; see also Hunt and Ivergård, 2007.)

Current-day understanding of organisation design implies participation by the operators of key processes in improving these processes. Operator participation is an important feature of control room operations. This process whereby someone identifies problems (as in AR) and is also empowered to resolve these problems is often called 'double looping' (Schön, 1982). It would be expected that operators take an active part in monitoring the control system and engage in creative development of the system and its work outputs. This is discussed in more detail in Chapter 11.

In Chapter 1 we discussed some general principles of the process for planning and building new industries. The main message is that the design process has no defined start and end points. The modern understanding of planning and design is that they are a continuous ongoing process. This is also the case for the design and development of control systems and related control rooms. Nowadays, control room operators are highly qualified and experienced personnel. Their skill and knowledge should obviously be used in an ongoing process of change and development of their own process industry.

2.4.3 Usage of Handbook Data

With reference to the above background, the operator will be allocated certain tasks and these will form the basis for determining the quantity and quality of personnel staffing levels. Factors such as educational background, practical experience, skills, personality variables, and choice of the technical principles, job aids, and information presentation and controllability would be used to assess these levels. Given this background, concrete 'knobs and dials' recommendations can be given on ergonomic grounds. This can be divided into the following sections:

1. General layout of information presentation and control devices.
2. Design of individual display devices and control devices (e.g., VDU screens, pointers, instruments, keyboards, knobs, buttons).
3. Design of workplaces, including work surfaces, rest surfaces, storage surfaces, secondary communications devices (e.g., telephones), and support in the form of lighting.
4. Room design.
5. General environmental design (e.g., climate, acoustics, air quality, windows).

On the first point, it is usual to start by determining the relative importance of the various tasks in the job from different viewpoints. One may attempt, for example, to define the information and the controls that are especially important and that require rapid attention or rapid corrective measures. This information and/or control device must thus be placed centrally and be clearly accessible.

The positioning of instruments and controls is usually determined by a cursory breakdown of the process into main functions (for example, well-defined process stages). Within these main functions, information/display devices are then positioned on the basis of *frequency of use* and *sequence of use*. In certain cases it may also be suitable to position the control and information devices on some form of function flow diagram. This type of function flow diagram is practically always required to complement the general overview of the process. In addition, some form of simplified technically orientated process model is also required, with signal lights for showing the current state of the process, perhaps also connected up to a general alarm panel.

The Risø group in Denmark mentions three main ways of presenting information (Rasmussen, 1979a; Goodstein, 1982):

1. Data/number presentation
2. Information connected with physical components in the process (e.g., tanks, pumps, reactors, condensers)
3. Functionally orientated information

Based on this classification, display systems have begun to be designed that are related to the functional presentation method (for example, Goodstein, 1982). The operator is thus working on a more abstract functional plane, and only when necessary will he or she go down to the more concrete and physical component level. Although this is an interesting approach, it can be questioned to some extent. It is absolutely correct to state that human beings build up abstract mental models (as discussed earlier in this chapter) of how an actual physical industrial process works. This abstract and mental model of the process does not require any direct agreement with the physical reality, but it is functional in the real meaning of the word. The operator needs the best available information about the actual system to enable him or her to build up a mental model quickly and effectively and to update it efficiently during continuous work. However, caution must be used when designing abstract models for the operator based on some form of theoretical working method. It is not certain that such models would help the operator to build and update his or her own mental process model more simply or more efficiently.

It is doubtful whether presenting this form of theoretically designed functional model to the operator is the right approach. The reason for doubt relates to the differences between individuals. Different operators work with widely differing models of how the process operates. Each individual operator probably builds up a personal model of a process on the basis of intellectual abilities, training, experience, and so on. What is typical for the model built up by an individual operator is that it suits his or her own ability to handle the system efficiently. In order to design an abstract functional model in advance to be presented directly to the operator, it is necessary to take into account the types of operators who will handle the system, and their abstract and functional model requirements. It is not certain, or even probable, that the functional models that are natural and suitable for system design are suitable for the operator, as system designers are probably accustomed to thinking in abstractions and will thus start from a completely different type of process model from that of most operators.

It was found in Ergolab's studies on the project of Computerisation in the Process Industries that the primary requirement of the operators was a set of simplified models of the physical design of the process (Ivergård, Istance, and Gunther, 1980). Therefore, to a large extent, the requirement was for information presentation more closely related to the types of jobs, which the operators had to do. If the operators were expected to work with more cognitively-orientated tasks—for example, for optimisation of operation—they would probably want a more functionally orientated model. If, on the other hand, the work was combined with maintenance work, for example, the physically-orientated models would be more suitable.

The following chapter describes detailed information on the design of information and control devices.

REFERENCES AND FURTHER READING

Akerblom, B. (1948). *Standing and Sitting Posture*. Stockholm: Nordiska Bokhandelns Förlag.
Bainbridge, E.A., and Beishon, J. (1964). The Place of Checklists in Ergonomic Job Analysis: Proceedings of the 2nd I.E.A. Congress. *Ergonomics* 7, 379–87.
Baker, C.H. (1963). *Vigilance*. New York: Ed Buchner and McGrath.
Bell, C.R., Provins, K.A., and Hiorns, R.W. (1964). Visual and Auditory Vigilance during Exposure to Hot and Humid Conditions. *Ergonomics* 7, 275–88.
Birmingham, H.P., and Taylor, F.V. (1961) A Design Philosophy for Man-Machine Control Systems. In H. Sinaiko (ed.), *Selected Papers on Human Factors in the Design and Use of Control Systems*, 67–87. New York: Dover.
Crossman, E.R.F.W. (1965). *Automation and Skill*. London: HMSO.
Damon, A., Stoudt, H. and McFarland, R. (1966). *The Human Body in Equipment Design*. Cambridge, MA: Harvard University Press.
Dickens, L., and Watkins, K. (1999). Action Research: Rethinking Lewin. *Management Learning* 30, 2 (June): 127–40.
Edwards, E. (1964). *Information Transmission*. London: Chapman and Hall.
Edwards, E., and Lees, F.P. (1973). *Man and Computer in Process Control*. London: Institute of Chemical Engineers.
Edwards, E., and Lees, F.P. (eds.). (1974). *The Human Operator in Process Control*. London: Taylor & Francis.
Goodstein, L.P. (1982). *An Integrated Display Set for Process Operators*. Roskilde, Denmark: Risø National Laboratory.
Greenwood, D., and Levin, M. (1998) *Introduction to Action Research: Social Research for Social Change*. Thousand Oaks, CA: Sage.
Hunt, Brian, and Ivergård, Toni. (2007). Organizational Climate and Workplace Efficiency. *Public Management Review* 9, 1, 27–47.
Ivergård, Toni. (1972). Check-out System for Self-service Shops, PhD Thesis, Loughborough University of Technology, Leicestershire.
Ivergård, T. (1980). *Utveckling och utformning av manöver-och kontrollrum*. Stockholm: Ergolab.

Ivergård, Toni. (1981). *Man-Computer Interaction in Public Systems: Advanced Study Institute on Man-Computer Interaction*. Amsterdam: Sijthort & Noordhoff International.

Ivergård, T. (1982). *Information Ergonomics*. Lund, Sweden: Studentlitteratur.

Ivergård, T., Istance, H., and Gunther, C. (1980). *Datorisering inom processindustrin*, Part 1. Stockholm: Ergolab.

Ivergård, Toni, and Hunt, Brian. (2007). A High Ceiling Environment for Learning and Creativity. Paper presented at the 4th ASIALICS Conference, Kuala Lumpur, Malaysia, 22nd–24th July.

Jones, I.C. (1967). The Design of Man-Machine Systems. *Ergonomics* 10, 101–11.

Lowe, E.I., and Hidden, A.E. (1973). *Computer Control in Industrial Processes*. London: Peter Peregrinus.

Martin, J. (1973). *Design of Man-Computer Dialogues*. Englewood Cliffs, NJ: Prentice-Hall.

McGrath, J.J. (1963). *Vigilance*. New York: Buchner and McBrath.

McNiff, Jean, and Whitehead, Jack. (2000). *Action Research in Organizations*. London: Routledge.

Meister, D. (1976). *Behavioral Foundations of System Development*. New York: John Wiley and Sons.

Raelin, J.A. (1997). A Model of Work-Based Learning. *Organizational Science* 8, 6 (December), 563–78.

Rasmussen, J. (1979a). *On the Structure of Knowledge—A Morphology of Mental Models in Man-Machine System Context*. Roskilde, Denmark: Risø National Laboratory.

Rasmussen, J. (1979b). Outlines of a Hybrid Model of the Process Plant Operator. In G. Salvendy (ed.), *Human-Computer Interaction*. Amsterdam: Elsevier.

Rasmussen, J. (1980). The Human Operator as a System Component. In H.T. Smith and T.R.G. Green (eds.), *Human Interaction with Computers*, 67–96. London: Academic Press.

Rasmussen, J., and Lind, M. (1982). *A Model of Human Decision Making in Complex Systems*. Roskilde, Denmark: Risø National Laboratory.

Rasmussen, J., and Rouse, W. R. (eds.). (1981). *Human Detection and Diagnosis of System Failures*. NATO Conference Series. New York: Plenum Press.

Salvendy, G., ed. (1986). *Advances in Human Factors/Ergonomics*, vols. 1–5. New York: Elsevier.

Schön, Donald. (1982). *The Reflective Practitioner: How Professionals Think in Action*. New York: Basic Books.

Shackel, Brian. (1969). Man-Computer Interaction. *Ergonomics* 12, 485–500.

Shackel, Brian (ed.). (1981). *Man-Computer Interaction: Human Factors Aspects of Computers and People*, no. 44. Maryland: Sijthoff and Noordhoff.

Shackel, B., and Whitfield, D. (1966). *Instruments and People*. London: HMSO.

Shahriani, Mohammed, Shee, Anirban, and Ortengren, Roland. (2006). The Development of Critical Criteria to Improve the Alarm System in a Process Industry. *Human Factors and Ergonomics in Manufacturing* 16, 3 (Summer), 321–37.

Sheridan, T.B., and Johannsen, G. (eds.). (1976). *Monitoring Behavior and Supervisory Control*. New York: Plenum Press.

Singleton, W.T. (1966). *Current Trends towards Systems Design*. Ergonomics for Industry, no. 12. London: Ministry of Technology.

Singleton, W.T. (1976). The Model Supervisor Dilemma. In T.B. Sheridan and G. Johannsen (eds.), *Monitoring Behavior and Supervisory Control*, 261–70. New York: Plenum Press.

Sorge, A., Hartmann, G., Warner, M., and Nicholas, I. (1982). *Microelectronics and Manpower in Manufacturing*. Berlin: International Institute of Management.

Tarrière, C., and Wisner, A. (1963). *The Road Vigilance Test, First Seminar on Continuous Work*. Leicestershire: Loughborough University of Technology.

Timonen, J. (1980). *Theoretic Modelling of the Human Process Operator*. Espo, Finland: Sähkötekniikan Laboratorio.

van Cott, H., and Kinkade, R.G. (1972). *Human Engineering Guide to Equipment Design*. Washington, DC: American Institute for Research.

Weiner, N. (1950). *The Human Use of Human Beings*. New York: Houghton Mifflin.

Welford, A.T. (1968). *Fundamentals of Skill*. London: Methuen Books.

Part II

Design of Information
and Control Devices

3 Design of Conventional Information Devices

Toni Ivergård and Brian Hunt

CONTENTS

3.1 INTRODUCTION

This chapter describes the design of traditional and conventional information devices, and also the design of devices for communication with computers. The design of instructions, forms, and tables will be dealt with towards the end of the chapter. In Chapter 4 we deal with the latest in design of information devices. A good understanding of the ergonomics of analogue information devices provides a firm foundation for a perceptual good design of the modern, more advanced, and flexible information devices. In other words, a good understanding of the design of analogue devices (instruments and controls) will be helpful in the design of the newer, more advanced

45

displays. Conventional instrumentation can easily be integrated into modern flexible displays using the latest forms of display technology. (See also Chapter 4.)

Also included in this chapter are instructions on how scales and scale markings on visual instruments should be designed, together with the advantages and disadvantages of different types of visual instruments. It appears, for example, that the common round meter with a moving pointer is best for most applications. Where more exact quantitative readings are necessary and there is plenty of time, the direct-reading digital instrument is best.

The chapter includes a relatively detailed specification for the design of visual display unit (VDU) screens. The main attribute of VDU screens is their flexibility, as they can be used for presenting many different forms of information. However, in control rooms, operators are often required to view and process a large amount of information simultaneously. Therefore it may be necessary to have several VDUs or to have access to other information devices such as overview displays as a complement to VDUs.

This chapter also describes methods for producing diagrams, codes, and symbols, and discusses various methods for using colour symbols. The use of colours may have some importance in simplifying the reading of process information. However, the use of colours should be limited, bearing in mind that a significant proportion of the population is colour-blind. Colours should be used to supply additional information so that the VDU can be read correctly even if all colour disappears.

In human/machine communication, the human operator receives information via the various information devices. In the control room, the information is either visual or auditory (and may be both), and can be either static or dynamic. For visually-impaired persons these information devices can also be tactile. Dynamic information is constantly changing, such as the information shown on speedometers, altimeters, radar, TV, temperature, and meters to measure pressure. Static information does not change over time and includes road markings, maps, notices, manuals, and any printed or written material.

In this chapter we deal primarily with the various types of dynamic information. Static information will also be covered to a certain extent, particularly in Section 3.3 (dealing with VDU screens) and also in Section 3.5. (Chapter 4 deals with the full set of new types of complex displays based on new and emerging technologies.)

Three main types of information devices will be covered in this section:

1. Traditional instruments
2. VDU screens
3. Sound signals

The new generation of VDUs (discussed in Chapter 4) represents—from a technological function—a dramatic paradigm shift in the area of visual displays. However, the functional principles discussed in Chapter 2 are also of relevance for the new generation of displays. The generic principles described in this chapter are of very large importance for a successful use of the new generation of displays. There are many great advantages with the new display technologies; for example, they make it possible for a good overview of the function of the system to be displayed.

This overview is easily lost in the use of cathode ray tube (CRT) and other small-scale display technologies.

Another great advantage of the new technology is its enormous flexibility. However, this also includes an apparent risk. The real strength of the new technology will only be realised if it is combined with the principles discussed in the current chapter. Technological advancement and inherent flexibility make it possible to simulate presentations which characterise the old 'classic' instruments. In turn, this makes the new technology more advanced from a functional and perceptual point of view. A classic example is the speedometer of a vehicle. A few decades ago vehicle manufacturers experimented with speedometers that displayed information in digital and thermometer-like forms. However, ergonomists and human-factor engineers in the automobile and aviation industries could rapidly prove that the superior display style was a traditional speedometer with a clock face and a pointer.

A further advantage is a combination of old analogue and the new digital technologies. This will create redundancy at the same time as it creates trust by the operator. In other words, in the detailed design of the application of the new display technology, we can use our knowledge from the classic display ergonomics.

3.2   TRADITIONAL INFORMATION DEVICES

Traditional instruments are still the most common form of information device in the control room. In modern control rooms, however, more and more information is being transferred to VDUs. Traditional instruments may be divided into the following subgroups with regard to their areas of use:

1. Instruments with associated control devices:
   Control regulation instruments—The instrument is read, and if necessary the operator adjusts the machine.
   Instruments for setting up—Instruments used for making changes in the running conditions.
   Instruments for following (tracking)—Instruments usually used in different types of vehicles (e.g., cars, planes).
   Instruments for indication—These show 'entrance', 'way ahead closed', 'backward', etc.
2. Instruments without control devices:
   Instruments for quantitative readings.
   Instruments for qualitative readings.
   Instruments for check readings—Instruments used for detecting and reporting a deviation from a normal value.
   Instruments for comparison—Used for checking that two machines are producing the same value.

Of the instruments above, the most common in industry are those for checking readings, for controlling regulation, and for setting up. The introduction of computer systems for process control, however, has meant that qualitative readings are becoming more common.

3.2.1 DIFFERENT TYPES OF VISUAL INSTRUMENTS

The choice from amongst the many different types of visual instruments available depends on the information to be presented and how it will be used. Figure 3.1 shows some of the more common types of instruments. Figure 3.1a shows a digital instrument which displays the various numbers directly. The instrument may have mechanically or electronically generated numbers. Figure 3.1b shows instruments with moving pointers and fixed scales, whereas in Figure 3.1c the instruments have fixed pointers and moving scales.

There are many variants of pointer instruments. Figure 3.2 gives examples of some of these:

1. Round and sector-shaped instruments (Figure 3.2a) are recommended for check and qualitative readings. Round ones are usually better than sector-shaped ones, but take up more room.
2. Vertical and horizontal scales with moving pointers (Figure 3.2b) are also good for check readings. They do not give as much information as that provided by the angle of the pointer in round or sector-shaped instruments. However, this type of instrument takes up little space.
3. Round and sector-shaped instruments with fixed pointers (Figure 3.2c) can be recommended where the whole scale does not need to be seen for quantitative readings with slow changes. Their design allows a relatively long scale to be used without taking up much panel space, but they may need a relatively large area behind the panel.
   a. Digital instrument
   b. Moving pointer, fixed scale

4	3	7	0	0

(a) Digital instrument

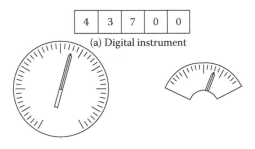

(b) Moving pointer, fixed scale

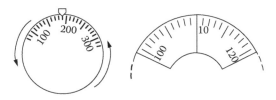

(c) Fixed pointer, moving scale

FIGURE 3.1   Various types of instruments.

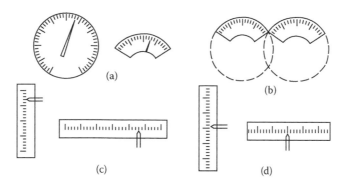

FIGURE 3.2   Varieties of instruments with moving pointers and moving scales.

5. Figure 3.2d shows vertical and horizontal instruments with fixed pointers. This type of instrument can also have a very long scale (see Figure 3.3). They cannot be recommended for cases other than where a very long scale is required.

Table 3.1 summarises the recommended areas of usage for different types of pointer instrument.

In certain cases, it may be desirable to choose other designs of instruments. One may wish to use the design to show the function of the instrument. Figure 3.4 gives examples of a round instrument with moving pointer indicating speed. The angle of the material or feature being controlled is shown on a sector-shaped instrument with moving pointer. The level is shown on a vertical instrument, and time changes are plotted out on paper with a moving pointer; the most recent time is marked at a suitable time interval.

3.2.2   DESIGN OF SCALES AND MARKINGS

The terms to be used in this section are defined as follows:

Scales range—The numerical difference between the lowest and the highest value on the scale.
Numbering interval—The numerical difference between two successive numbers on the scale.
Scale marking interval—The numerical difference between two scale markings.

Before choosing a scale for an instrument, one must determine the precision required. Figure 3.5 gives examples of different precisions. It is generally true that the least exact scale that fulfils system requirements should be chosen, thus avoiding unnecessary accuracy. If possible, the information should be given in units that require no modification by the operator before they can be used. An example of this might be percentage figures being used instead of the actual number of revolutions,

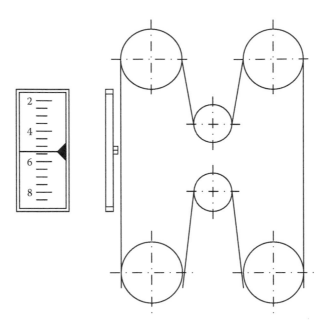

FIGURE 3.3   An instrument that covers a large-scale range can be made with a long moving scale and a fixed pointer.

so that the operator need not remember different top speeds on different machines (a scale speed is often the same percentage of top speed on all machines). The percentage scale also gives fewer digits to read off (see Figure 3.6).

Figure 3.7 shows examples of recommended scale designs for instruments with different scale ranges. On the horizontal axis are the marking intervals, 1, 2, 5 (and 1/10 and 10 times these values). This means that for 2 there are two (or 0.2 or 20) units between each subsidiary marking. On the vertical axis are the numbering intervals, that is, the numerical difference between two successive numbers on the scale.

The matrix in Figure 3.7 shows examples of recommended scales and examples of how the markings can be designed. The scale can either be marked with large, medium, or small sizes of marks. Certain of the scales use all three sizes. Large markings are always used for the numbers of the scale. Medium-sized and small markings are used for subdivisions, that is, those divisions which lie between the numbered markings.

There must not be more than nine subdivisions between numerical markings. The scales should preferably be designed so that interpolation between divisions is not necessary. Where there is a lack of space, however, it is better to allow interpolation than to clutter up the instrument with too many markings. If there is room for the scale illustrated in Figure 3.8a, this is the one to choose; otherwise, choose the one shown in Figure 3.8c. The scale in Figure 3.8b is not suitable. If the alternative in Figure 3.8c is chosen and the scale has to be read off to the nearest whole number, this means having to interpolate one step between markings. The smallest

TABLE 3.1

Recommended Areas of Use for Different Types of Pointer Instruments

	A — Moving Pointer			B1 — Fixed Pointer			B2 — Window		C	D	E
	Round	Linear Horizontal	Linear Vertical	Round	Linear Horizontal	Linear Vertical	Round	Linear	Counter	Switch	Lamps
Without Controls:											
Quantitative reading, slow change	o	o	o	o	o	o	o	o	xx	o	o
Quantitative reading, fast change	x	x	o	o	o	o	(x)	(x)	o	o	o
Quantitative reading, direction	(x)	o	x	o	o	o	o	o	o	o	o
Quantitative reading, speed	x	o	o	(x)	o	o	o	o	o	o	o
Control/check reading	xx	(x)	(x)	o	o	o	o	o	o	o	o
Comparison, fast	xx	o	o	o	o	o	o	o	o	o	o
Comparison, slow	(x)	o	o	o	o	o	o	o	xx	o	o
Warning	(x)	o	o	o	o	o	o	o	o	x	xx
With Controls:											
Control/check adjustment	xx	(x)	o	o	o	o	o	o	o	o	o
Setting up	x	x	(x)	o	o	o			o		
(Tracking)				Mostly Special Instrument							
Indicating	xx	o	o	o	o	o	o	o	o	x	x

Legend: xx, very good; x, good; (x), uncertain; o, unsuitable.

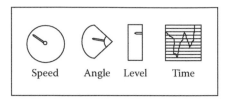

FIGURE 3.4   Form coding through the shape of the instrument.

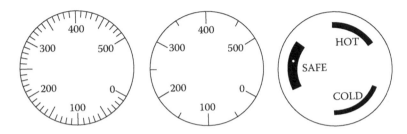

FIGURE 3.5   Different scale complexities.

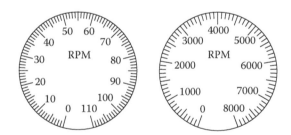

FIGURE 3.6   Comparison of percentage (left) and absolute value (right) scales.

step for which it is necessary to interpolate is called the *subjective scale division*. There should be zero, two, or five subjective scale divisions for every marked interval. The basic factor in the reliability of reading an instrument dial is the physical width (measured as an angle of view) of subjective scale divisions (that is, zero, two, or five subjective scale divisions per marked interval). The scale is subdivided in such a way that the reading tolerance and accuracy one wishes the instrument to have are achieved. Accuracy must not be greater or even less than the allowable pointer error for the instrument in question. If it is necessary to interpolate fifths in order for the correct degree of accuracy to be obtained, the angle of view must be five times the angle of the subjective scale division, which equals the smallest division for which it is necessary to interpolate. If the viewing angle for the subjective scale division is less than a certain critical size, there will be considerable errors in

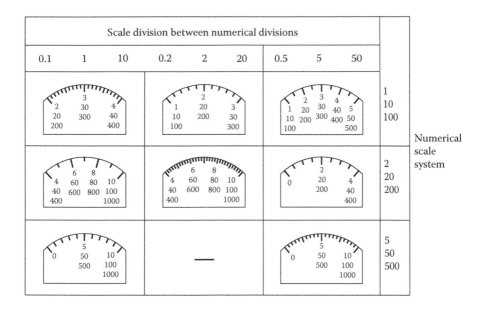

Scale division between numerical divisions									Numerical scale system
0.1	1	10	0.2	2	20	0.5	5	50	

FIGURE 3.7 Examples on recommended scale division for different sizes of the scale.

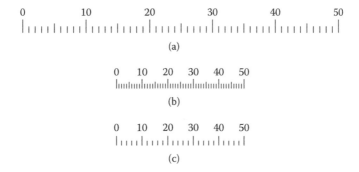

FIGURE 3.8 The physical size of the scale and the number of scale divisions. Scale (b) is not to be recommended.

reading it. If, on the other hand, the angle of view of a subjective scale division is greater than this critical size, one cannot expect any further increase in reliability of reading, as shown in Figure 3.9, according to which the width of the critical scale division is 2 minutes of arc.

The reading reliability over this critical value would be expected to rise to 99% for people with long experience of instrument reading. When designing a scale, therefore, one would start by determining the greatest distance at which the scale is going to have to be read. Based on the distance and the critical angle of 2 minutes of arc, the smallest size of the subjective scale divisions which will be needed on a scale is calculated from the total measuring range the instrument is to have. Knowing that

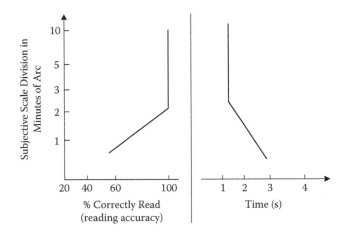

FIGURE 3.9   Effect of subjective scale division on reading accuracy and time. Increasing the subjective scale division to greater than 2 minutes of arc does not increase the reading accuracy or reduce the reading time. (Murrell, 1958. With permission.)

there should be two or five subjective scale divisions between each marked interval, the number of marked intervals can be selected. There should be two, four, or five marked intervals for each numbered interval.

For a scale marked in five subjective scale divisions per marked interval, and which has a total of 100 subjective scale intervals, the total length of the scale is determined according to the reading distance obtained from the following formula:

$$D = 14.4 \, L \tag{3.1}$$

D is the reading distance and L is the length of the scale. D and L have the same units of length, for example, cm. If one does not have such a 'standard scale' with the above relationship between subjective scale division and marking interval, there is a correction factor that may be added to the formula:

$$14{\cdot}4 \times L = [D \times (i \times n)] \, / \, 100 \tag{3.2}$$

where i is the number of subjective scale divisions into which each marking interval is to be interpolated, and n is the number of marking intervals. In practice one is often tied to existing standard instruments. Here one can instead evaluate the maximum reading distance for the different instruments and determine whether this is sufficiently far for the actual application.

The scale itself should not have long lines joining up the scale markings, but should be open (see Figure 3.10). Figure 3.11 gives the recommended measurements of the scale marking for (a) normal visual conditions, and (b) low light levels. In this case, the reading distance is approximately 70 cm. The measurements given in

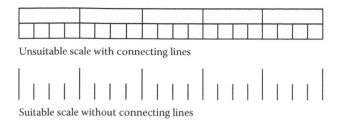

Unsuitable scale with connecting lines

Suitable scale without connecting lines

FIGURE 3.10 Examples of suitable scale designs.

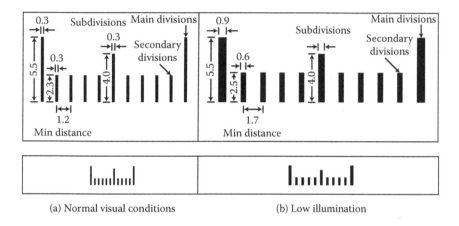

(a) Normal visual conditions   (b) Low illumination

FIGURE 3.11 Recommended design of scales (mm).

Figure 3.11 apply not only to straight instruments but also to circular and segment-shaped ones.

The design of the letters and numbers on the scale is important for giving the best possible reading reliability. The following points should be noted especially:

1. The ratio between the thickness of the line and the height of the letter/number for white numbers on a black background should be between 1:10 and 1:20. If the background is dark and the numbers are illuminated from behind, the optimum is between 1:10 and 1:40. The optimum for black letters on a white background lies between 1:6 and 1:8. Black lettering on a white background is normally preferable. For dark-adapted and low-illumination conditions, white numbers on a black background are preferable (for night viewing in combination with red instrument lighting, the white numbers will be seen as red).
2. The ratio between the height of the numbers/letters and their width should lie between 2:1 and 0.7:1. The optimum is 1.25:1.

3. The appearance of the numbers/letters is important. As a guide, the numbers/letters should be simple in their design. Serifs and too many lines complicate the reading. However, they must not be so simple as to become difficult to understand.
4. Letters/numbers must not be positioned on the scale in such a way that they are shaded by the pointer, and they must be vertical.

The following recommendations are made on the design of the pointer:

1. The pointer must be long enough to reach but not to overlap the scale markings.
2. The pointer should lie as close to the surface of the dial as possible in order to avoid parallax errors.
3. The section of the pointer that goes from the pivot to the scale markings should be of the same colour as the scale markings. The remaining part of the pointer should be of the same colour as the dial face.
4. On horizontal scales, the pointer should sit on the top of the scale, and for vertical scales it should be on the right-hand side with the numbers on the left.

3.2.3   SOUND SIGNALS

Certain types of information are better transmitted as sound signals rather than visuals. This is the case in the following situations:

1. When the signal is originally acoustic.
2. Warning signals. The advantage is that the operator does not need to see the signal in order to detect it; that is, he does not need to look at the instrument constantly.
3. Where the operator lacks training and experience of coded messages.
4. Where two-way communication is required.
5. Where the message concerns something that will occur in the future, e.g., the countdown to the start-up of a process.
6. In stressful situations, where there is a possibility that the operator would forget what a coded message meant.

Tones are preferable in the following situations:

1. For simplicity.
2. Where the operator is trained to understand coded messages.
3. Where rapid action is required.
4. In situations where it is difficult to hear speech (tones can be heard in situations where speech is inaudible).
5. Where it is undesirable or unnecessary for others to understand the message.
6. If the operator's job involves constant talking.
7. In those cases where speech could interfere with other speech messages.

Warning signals are without doubt the most common form of sound information. Sound is an information source that is little used, with the exception of warning signals. Sound signals/information could probably be used to advantage in giving spoken instructions in disaster and other acute situations, for example, (1) that there is a fire at a particular location, and (2) what action should be taken. In this type of situation one can use a calm voice to give further instructions on necessary actions, such as clear and simple instructions on how to evacuate the building. Various forms of acoustic alarms should be able to be complemented with tones that would give preliminary information on the type of fault. If there is a hierarchy of alarms, different tones can be used for different groups of alarms. If detailed information is required about the alarms, questions can be entered via the keyboard or directly using the voice. Section 3.6 includes a description of the design of computer-generated speech and computers that understand speech.

3.3   VISUAL DISPLAY UNITS (VDUS)

In the past the information devices traditionally used in control rooms normally had only a single area of application. The instrument was normally electrically connected directly to some form of measuring device. In certain cases switches were used, which allowed the same instrument to be connected directly to several measuring devices. For example, this was usually the case for temperature readings where it was common to have alarms connected to all the measuring points. One then had to turn a special switch in order to be able to read the temperature at all the measuring points.

In the modern computerised process control system, information is collected from many measuring points in the system. Information is stored in computers and subjected to further processes, which then pass on the information to various receivers, such as the different information devices in the control room. The information devices connected to the computer system are normally the type that can simultaneously carry out several functions, that is, which can receive different forms of information, such as cathode ray tubes (CRTs). This form of flexible display could formally be described in Chapter 4, but for historical reasons we have retained it in the current chapter.

3.3.1   VDU DESIGN

A visual display unit is the collective term for information devices designed for several different purposes (for example, for reading off temperature, shaft speed, instruction, information on handbook data). There are many different types of VDUs, of which the most common is the CRT. But there are many other types, such as liquid crystals, plasma displays, and matrices built up of different types of lamps or light-emitting diodes (LEDs).

The important advantage of the VDU is that it can handle most types of presentations, although the various technical solutions do have some limitations, for example, the shape of characters that can be formed on the screen. The pictorial alternative is provided (within the framework of the technical specifications) through programming of the computer. It must be stressed, however, that the programming must suit the actual process operator and the nature of the work and not just reflect how the programmer feels that interactive VDUs should be designed. Unfortunately, the latter is often the case. This means that many VDU applications in industry have shortcomings or are simply unsuitable and faulty; this does not only mean discomfort and limitations for the operator, but it also gives rise to poor performance and an unnecessary number of errors.

There now follow some guidelines on the design of information to be displayed on VDU screens. These recommendations are not the final ones, which would be applicable to all the different applications and control situations. Instead, they should be treated as a form of checklist, which can be used in the design of human-computer communication systems in the control room for process control. It is also important to emphasise that it is virtually impossible to choose the technical components first and then expect to get a functional solution. Even if ergonomic requirements are specified for each individual component, there is a risk that the overall solution will be bad. Instead, one must work in accordance with the systematic method described in Chapter 13. Technical solutions for the design of the interface with the operator can only be produced from detailed descriptions of the job content and analyses of the various work and skill requirements.

Summarised below is equipment that can be used for different forms of information presentation, together with some advantages and disadvantages.

1. Presentation of set value of a variable or deviation from set value:
   Points or a bar with a scale (e.g., a thermometer)
   Traditional instrument, plasma, or LED forming a bar or equivalent on CRT
   Moving numbers
   Dot matrix, seven-segment plasma panel, CRT, etc.
2. Trend or time history of a variable:
   Graphics
   Matrix printer or plotter
   Line diagram:
   Alternative 1: Line printer or plotter
   Alternative 2: CRT
3. Relationship between set values of several different values.
   Line or bar diagram:
   Alternative 1: CRT or plasma panel
   Alternative 2: Multi-pen printer or plotter
4. The way a circuit works.
   Illuminated indicators or a displayed mnemonic:
   Alternative 1: Lit-up buttons, LEDs, rear-projected displays
   Alternative 2: CRT
   Alternative 3: Printer
5. Alarms
   Matrix:
   Alternative 1: Signal/number board
   Alternative 2: Special reserved CRT

Tables in chronological, hierarchical, or random order:
Alternative 1: CRT reserved for the job
Alternative 2: Line printer
6. Text
Lit points in certain groupings:
Alternative 1: Plasma grouping (usually a 5 x 7 point matrix per symbol)
Alternative 2: CRT at 25 Hz with interlacing
Alternative 3: CRT at 50 Hz frame frequency with no interlacing
Illuminated line segments:
CRT with DC positioning (*x/y* techniques)
7. Diagrams
Lit points in a matrix:
Alternative 1: Plasma grouping
Alternative 2: CRT with 25 Hz frame frequency and interlacing
Illuminated line segments:
CRT with DC positioning

For the *presentation of the set value*, most types of equipment have advantages and disadvantages. The advantage of pointers or thermometer-type bars is that they give a certain indication of the size of the changes as they occur. It is more difficult to display this feature with a digital presentation. The advantage of a digital presentation is that it is easy to present a large amount of data at the same time. On the other hand, a numerical presentation gives a poor understanding of size and quantity. This also makes it difficult to compare several values at the same time.

The ordinary printer or plotter is good for *trends and time histories* of values. However, it is difficult to print out several values at the same time, although good documentation (copies) is provided. The bar chart may mean more difficulty with, for example, paper handling, and also require some form of identification of the individual diagrams. CRTs produce a somewhat poorer picture than graphic printers, but on the other hand are more flexible. It is also easy to use a CRT to display different variables in turn. No hard copy can be produced directly from the screen, but this can be obtained using cameras or additional equipment.

In order to present the *relationship between variables and set values*, CRTs and plasma panels are flexible and suitable in many cases. However, resolution and sharpness are relatively poor in comparison with a multi-pen plotter. The disadvantage of the multi-pen plotter is that the format is often limited. Due to the lack of resolution, the same is also true for the CRT.

In order to see how different *circuits are working*, lit press-buttons are a good system but impractical if many functions have to be presented at the same time. The CRT can be tabulated for a number of different circuits and functions, and can also be used in conjunction with cursors and light pens. With the more traditional old type of display for *alarms*, the greatest disadvantage is a lack of flexibility. These also need a lot of space, but do give a good overview and continuous accessibility. The system also works well interactively if the diagram has illuminated press-buttons on it, where a function of the buttons is for acknowledgement or corrective actions.

The great advantage with using a CRT is that it is easy to update. It is important, however, to have a CRT reserved specifically for this purpose. In addition, the system is compact and easy to use interactively with a light pen, for example.

If the alarms are presented in tabular form with text, chronologically, hierarchically, or in random order (instead of in a matrix), the greatest disadvantage is that the overview becomes lost. On the other hand, it is possible to include an extra line along with the alarm to advise the action that should be taken. Compared with the CRT, the printer also has the advantage that it produces documentation of the alarm. This information can also be stored in a computer. However, from a documentation point of view, it might also be useful to have paper copies.

The plasma and similar displays have a number of advantages in the *text presentation*. For example, they are very thin, don't cause any flickering effect, and the surface is flat. The contrast on the light surfaces can be high, and it is easy to programme.

If the recommendations given in the next section are followed, the risk of visual difficulties and eye fatigue will be considerably reduced. Problems may remain, however, for people with vision defects. The operator must then obtain spectacles, which are made and tested specifically for work with CRTs.

3.3.2   DESIGN OF CATHODE RAY TUBES

The following section presents some brief recommendations on the requirements, which should be placed on CRTs, currently the most common form of information presentation aid in the computer control system. The basic design features of a CRT are that the information shall be visible and easy to read. It is important to choose a suitable coding and presentation of information for particular applications. These points are dealt with in Section 3.4. The following recommendations assume a screen with a dark background and light text. This form of CRT demands very careful planning of the lighting (see Chapter 7). There is much to suggest that screens with dark text on a light background give a better result if one can avoid the problems of flicker, which easily occur on light screens (Berns and Herring, 1985).

Some approximate guidelines are given below:

1. Clarity
   a. *Luminance*—Minimum 85 cd/m² for characters (250 asb),* with an optimum of 171 cd/m² (500 asb).
   b. *Contrast*—The optimal character contrast is 94%, but down to 90% is acceptable.
   c. *Light/dark characters*—If flickering can be avoided, dark characters on a light background are preferable. Otherwise, and if other environmental conditions are suitable, light characters on a dark background can be accepted.
   d. *Flickering*—For the optimal luminance level of 171 cd/m², a frame regeneration rate of at least 50 Hz is required. Certain types of phosphor (e.g., p-20) may need higher frame rates, up to and even over 60 Hz (frequencies of 60 Hz may interfere with mains electrical frequencies of 50 Hz in Europe and other places).

---

* 1 asb (apostilb) is the luminance of a white surface illuminated by 1 lux.

2. Legibility

    a. *Character size*—The height varies between 16 and 27 minutes of arc (visual angle), a width-to-height ratio of 0.75, and a ratio of height to thickness of the strokes making up the characters of between 10:1 and 6:1.

    b. *Character shape*—A dot matrix using 9 x 7 points produces the best symbols. There are many different ways of designing letters and numbers and it is not always clear which is the best. It is, however, important to choose types that do not cause confusion between letters and numbers, as in O and Q, T and Y, S and 5, I with 2 and K, I and 1, O with 8 and 0.

    c. *Resolution*—Ten raster lines per character height and more for nonalphanumeric characters.

    d. *Character separation*—50% of character height and 100% between lines.

    e. *Reading distance*—For characters of 3.2 mm in height, a reading distance of 69 cm is acceptable. In most practical applications, the reading distance is determined by the height of the characters. The screen should be read from directly in front; angles of more than 30° from the optimum considerably reduce legibility.

    f. *Colour*—Colours that lie well outside the optimum visual spectrum must be avoided, particularly those in the blue ultraviolet (UV) region. Colours in the yellow-green range are good.

    g. *Flashing* (e.g., to gain attention)—A 3-Hz flashing frequency has no adverse effect on legibility.

    h. *Cursors*—A rectangular cursor with a 3-Hz flashing rate is thought to be best.

3. Coding

Coding improves *performance* considerably, especially for simple information processing tasks.

    a. *Colour coding*—Best for localisation, calculation, and comparison of different information.

    b. *Alphanumeric codes*—For recognition and identification of information, these are the best. Colour coding is the next best. Coding is of little or no help in *quantifying* or *size estimation*.

    c. *Components*—A code should not comprise more than seven different components (e.g., seven letters or numbers). The fewer components the code uses, the better. The *efficiency* of the coding must be seen in relation to the total construction of pictures on the screen.

    d. *Code groups*—Different groups of codes, each consisting of components, should, if possible, not contain the same components.

4. Construction of Picture on the Screen

The basic issue in the determination and design of the picture is the function that is to be fulfilled. The following information may be used to provide certain guidelines (see also Section 3.3):

    a. Digital presentation is best for quantitative information.

    b.   Pointers on circular scales are best for showing changes.

    c.   Vertical and circular scales with moving pointers are good for check readings.

    d.   Chart recorders are valuable for seeing instrument faults and for obtaining a rapid impression of the system's response.

Diagrams are often preferable to graphs. Histograms are especially easy to read. Curves, however, are preferable for reading trends. Several curves can be presented to advantage for comparisons between several variables. Here, semigraphic screens are better, but x/y screens are to be preferred above all.

3.3.3　Design of Tables

- Whether data should be presented in columns or rows depends on what is more natural for the task in question.
- Letters are better for identification than numbers.
- Numbers should be arranged in as few columns as possible.

The design of pictures and the positioning of text within each frame is usually a compromise between several different factors. It is usually very difficult to adhere strictly to the various criteria that are set in designing different frames.

The design of frames or pictures is also determined by the length of the text, as there is only a limited area available; if the text is very long, its length will determine its positioning. In such cases it may be necessary to abbreviate the text. When doing this, it is of the utmost importance to maintain as far as possible the comprehensibility of a word even though it is abbreviated.

Through the placing of the same information in the same position in different frames, the possibility of errors by the operators is reduced, together with search time. This is particularly true for beginners and those who use the terminal infrequently. Consistent positioning of error messages is also very important as they must always be immediately obvious to the operator. Messages may be overlooked if the operator has to search for them.

The most fundamental criteria by which information to be used in a frame should be analysed are:

1. The importance of the information.
2. Its frequency of use.
3. The sequence of use.

It is clear that even though these criteria have importance on their own, they are very closely connected. One example of this is the first information to come up on the screen, which is often both the most important and the most frequently used information.

When the principles for the design of the pictures and the text have been determined, the information should be presented in such a way as to simplify the task of the operator as much as possible.

Formerly, the CRT type of display used to have poor picture sharpness (focus) at the perimeter caused by the curve on the screen edge, which reduced the visual clarity of the characters. On many screens, this may mean reducing the amount of information on both the left- and the right-hand sides of the screen by three, four, or even five columns in order to avoid this poor focus area. The effective area of the screen is thus reduced. Another aspect of the design is whether to use every line or every alternate line. No general answer can be given to this, as it depends partly on the design of the screen (that is, the distance between the lines) and partly on the amount of information on the screen. Apparently, the modern generation of CRT has solved most of these problems. However, the continued usage for 24 hours per day, 365 days per year, does create problems for all types of displays and the user.

After determining the above criteria comes the question of how tightly together the information should be positioned. The amount of information may be defined as the quantity of information within a defined area. If too much information is displayed, operator performance is decreased (Stewart, 1976; Cakir, Hart, and Stewart, 1979). Stewart suggests that the search time in seconds is roughly one-fifth of the number of alternatives to be searched through. Even if a very experienced operator is able to be selective in searching for information, irrelevant information means that the search time and error frequency will increase, thus reducing the performance of the operator and the system.

3.3.4   Advantages and Disadvantages of VDUs

In older types of control rooms, every variable was represented by its own instrument, regulator, or control device. These were placed together on a panel. Over the years, various ergonomic rules for the design of panels have been produced with the aim of achieving the best possible operational conditions. The development of VDUs and computers has now provided new forms of information presentation, monitoring, and controlling. Research by Ergolab has shown that many operators prefer the conventional instrumentation over the computerised versions (Ivergård, Istance, and Günther, 1980), which are often described as being akin to monitoring the process through a keyhole.

Industrial researchers have described how people scan a picture for the object they are seeking (Stark, 1983). It appears that the normal strategy is to fix the gaze on a certain number of points in order to identify the picture; the points chosen on which to set the sight depend on the object. Stark (1983) also found that the experienced viewer will miss some of the points previously fixated on. Despite jumping over them, the viewer will still be able to make the correct interpretation of the picture. In addition, the viewer will be able to detect whether there are any changes in the parts of the picture jumped over. In other words, people use their peripheral vision for checking whether there are any changes in the picture. Computer image-recognition programmes operate in a similar fashion by fixating on specific points in the image, such as the face or the body. It is likely that a process control room operator will work in a similar fashion in the monitoring of a VDU or a conventional panel. The operator will fixate on a number of points in order to obtain a picture of the current situation in the process. The operator then updates his or her mental model of the status of the process. The experienced operator has a considerably higher performance level in monitoring, and probably fixates on considerably fewer points to estimate the status of the whole process.

Sometimes the operator only has access to a number of images presented on VDU screens. This does not allow parts of the process to be updated, other than those currently presented on the screen. In more conventional instrumentation, which presents all of the information in parallel, the operator has very different possibilities. By fixating the gaze actively on certain parts, and using the peripheral vision for other parts, the operator can continuously update his or her mental model of the status of the whole process.

Given the above background information, it can be seen that there is a natural division of the viewing process into two types: active and passive vision. When someone fixes his or her gaze on one or a certain number of points in order to identify an object, this is active vision. In parallel with the active process, passive vision is occurring via the more peripheral parts of the retina.

In active vision, the gaze is turned towards the object and fixated on the central part of the retina where the cones are most dense, that is, the fovea. In the area around the fovea, the rods dominate. The rods are considerably more sensitive to light and can therefore work under relatively dark conditions. They are also sensitive to movement and changes and are therefore used in pattern recognition. The cones, on the other hand, need more light. The ability to distinguish detail, mainly by the cones, is also thought to increase in proportion to an increase in the light level. There are also cones that have the ability to distinguish colours. This allows us to draw certain conclusions regarding active and passive vision. Active vision allows the identification of colours and small objects under good lighting conditions. Passive vision permits recognition of patterns, and therefore especially changes and movements. Passive vision also works if the object is in motion.

In control room work, active vision is used to make detailed readings of a more quantitative nature. It is also used in the identification of colour codes and in the detection of small differences, in curves or diagrams, for example. Active vision is excellent for VDU viewing. There the operator can call up a particular frame and adjust such parameters as the set value.

Using passive vision, one could tour the control room and identify changes in the process pattern—for example, lights being lit or extinguished on a panel. Schematic representations of a process with built-in indicators (for example, signal lamps indicating deviations) are a suitable type of presentation for passive vision. In other words, passive vision is perhaps best applied on a more traditional type of instrument panel.

The choice of presentation method depends very much on the task of the operator. There are three main rationales for having an operator in the control room:

1. The operator acts as a supervisor in order to carry out certain standardised routine tasks, which for various reasons, have not been automated. The operator also has the task of calling for expert help when some unforeseen incident occurs.
2. The operator is a qualified expert with the job of carrying out production planning and optimisation tasks. Simpler, more routine, and predictable

types of faults are dealt with by the operator. The more serious, unforeseen faults are passed on to special maintenance experts.

3. The operator is primarily a maintenance-orientated expert who gets information on the process from other sources, especially those that are economic in character. The operator usually looks after production quality matters himself. The operator is expected to be able to deal with most unforeseen and difficult faults and events in the process.

If the first of these alternatives is chosen, one can determine relatively accurately beforehand what type of information the operator will need in different situations. There is less of a requirement for the more detailed type of overview information. The conventional type of instrumentation therefore provides very little information to this operator and he or she can largely rely on a number of VDU screens with a predetermined programme of frames.

The expert operator who is either production-orientated or maintenance-orientated has a considerably greater need for more detailed, continuous, and parallel presentation of the whole process. If the information is presented on VDUs, most processes would require a large number of VDU screens or extremely large VDU screens. The alternative is to require the operator, even during normal running conditions, to sit down and leaf through all the status frames. An operator would have to do this in order to update himself actively on the process status and to build up his knowledge of the functioning of the process. Instrumentation of the conventional type offers completely different possibilities for the operator to update his mental picture of the process. This can be done by looking at the instrumentation both consciously and unconsciously.

The production-orientated operator needs to be able to see a relatively detailed and dynamic functionally-orientated process model. The maintenance-orientated operator, on the other hand, needs to have a more physically-orientated model. The traditional instrument and control panel is a good alternative, but the conventional type of instrumentation is not preferable to a VDU in all processes.

The VDU screen has a great advantage in being flexible. Colour monitors with high resolution and detailed pictures also allow the presentation of a large quantity of information in a limited workspace. VDUs are also suitable for presenting different types of information. This may be graphical information (for example, maps of temperature distributions [isotherms], pressure distributions [isobars], and so forth). Even if VDUs cannot always wholly replace conventional instrumentation, they are a necessary complement in modern process control.

Table 3.2 gives a comparison between conventional instrumentation and electronic VDUs. It may be seen that the VDU has many advantageous characteristics and that the disadvantages are largely of an ergonomic nature. In addition to the characteristics given in Table 3.2, visual problems should be considered. These almost always occur when using the CRT-type of visual display. In practice, both VDUs and more conventional instrumentation are required in most cases. When modernising an existing control room, it is best to keep the old instrumentation as a reserve and a complement to the VDUs. When building from new, it is rarely sensible to install both the conventional type of instrumentation and VDUs. On the other hand, it may

TABLE 3.2

Characteristics of Conventional and Electronic Display Devices

Attribute	Conventional	Electronic/Advanced
Nature of presentation	Parallel	Serial
Mode of presentation (digital, analogue, etc.)	Fixed	Variable
Availability of information	Continuous	On request
Relationships among items of information	Static	Dynamic
Nature of interface	Inflexible	Flexible
Incorporation of diagnostic aids	Limited	Readily feasible
Ability to modify or update	Difficult	Easy
Redundancy of displays about control rooms	Costly	Relatively inexpensive
Control room size requirements	Relatively large	Relatively compact
Compatibility between process response and control movement	High	None
Control display feedback	High	None
Nature of overview information	Detailed	Summarised
Visual conditions	Reflected	Flickering, radiated light

Source: Developed from Seminara (1980).

be a useful complement to the VDUs to produce some form of detailed overview panel. This can schematically and dynamically describe the physical design of the process. It is often desirable, and sometimes necessary, to provide dynamic information on the overview panel. This may, for example, show which valves are open or closed, which pumps are working or not. It may sometimes even be desirable to provide the overview board with quantitative information such as flow rates and levels. The quantity of dynamic information provided on the panel depends very much on the type of process. In a continuous, stable process, with relatively few starts and stops, a static overview panel is often sufficient. These may be described as processes with high inertia, where changes in the process variables take place over many minutes, and where a detailed diagram of the principles showing the flow and the interconnections is required. In addition, the most important physical units must be marked on the diagram. A good complement to this form of presentation may be a three-dimensional model of the process.

A dynamic and detailed overview board is required where a complicated network with several alternative connection routes exists. The connections in one part of the network will affect the conditions in another part of the network. A typical example is an electricity supply network, and to a lesser extent, a water distribution network. A relatively complex batch process also needs a detailed dynamic overview panel. If the batch process is simple enough to be presented as a picture on a VDU screen, an overview panel is not necessary. An overview panel is also required on a continuous process with many stops and starts, and where the course of events is relatively rapid. Such processes would have changes occurring within seconds or, at the most, a few minutes. Production-orientated operators will always need a functional overview diagram. This can be suitably presented on a VDU screen. The maintenance-orientated operator has a special need for detail in the overview panel.

3.4   INSTRUCTIONS, FORMS, TABLES, AND CODES

Instructions, notices, and forms give printed information to the operator, often in connection with a product. It is of the utmost importance that this information is presented clearly and concisely and that it is easily read. This is particularly the case when an operator is faced with a new product, one that he rarely uses, or in a situation where time is limited.

The following rules apply:

1. *Letters*—Capitals alone are recommended for general use, but a combination of capitals and lowercase (small) letters are allowed (first letter a capital followed by lowercase letters). The ratio between width and height of the uprights should be between 1:6 and 1:8. The ratio of width to height of letters should be 3:5.
2. *Short messages*—In order to save time and space, the text should be as short as possible, whilst still retaining clarity.
3. *Choice of words*—Words and meanings should be kept as simple as possible (see above). Only well-known words should be used. For special populations of people, one may use common technical terms, for example, the use of aeronautical terms when addressing pilots.
4. *Clarity*—Instructions must be short and easily understood. One very simple way to test this is to try out different instructions on different people.
5. *Contrast*—In work where dark vision is required, letters should be white or light yellow on a black or other dark (e.g., brown) background. Where no dark vision is required, the letters should be black on a matt white background. Other colours may also be used for coding purposes, but they should always be chosen with maximum contrast in mind.
6. *Size*—The recommended size of letters depends on the reading distance. At a distance of 70 cm or less, in low lighting conditions, the size should be about 5 mm. In good light this can be reduced to 2.5 mm.

3.4.1   DIAGRAMS AND TABLES

The following recommendations apply to diagrams and tables:

1. Diagrams are always better than tables if the shape, variation, or connection between materials is of interest or if interpolation is necessary. If not, tables are preferable.

2. Simplify the table as much as possible without reducing its accuracy and without the need for interpolations (an example of this would be where, if the depth markings on a water tank level are not sufficient, several calculations are required to interpret water level).

3. Leave at least 4 to 5 mm between columns, which are not separated by vertical lines.
4. Where the table columns are long (more than six lines), they should be divided into groups of three or four lines. Leave some space between each group.
5. Diagrams should be drawn so that the numbered axis lines are darker than the unnumbered ones. Where only every tenth line is numbered, the fifth line should also be denser than the others, but less dense than the numbered ones.
6. Avoid combining too many parameters in the same diagram; there should be no more than three. If more parameters are shown, more diagrams must be used.

3.4.2   CODES AND SYMBOLS

Different sorts of traditional symbols are used by humans to convey a particular meaning; for example, the cross represents Christianity for a large proportion of humanity. The alphabet can be used to build up different definitive meanings using various systematic rules. It is important in the construction of an artificial language that all the users agree on the rules that are to be applied to it. This is the case both in more complex languages (for example, those that resemble the natural ones) and in the simpler ones, such as those used by machines of different types or as explanations on VDU screens.

The assumptions necessary before a language can be built up are, first, that everyone agrees what the different characters mean, and secondly that there are certain predetermined semantics. There must also be a grammar, which specifies the way in which the different symbols can be combined. Finally, the symbols must be designed in such a way that they can be understood by other people. As far as visual symbols are concerned, they must be able to be clearly seen and understood in the situation in which they are designed for use.

Languages may be termed *natural* or *artificial*. Natural languages are often learned in childhood, or after very comprehensive training and/or education. More complex artificial languages that are to be used in a similar way to the natural languages also require a very comprehensive education and training. In the design of languages for use in different technical applications in industry, special stress must be laid on the need to make the rules simple so that as far as possible they are self-evident. Even if one can set out such very simple rules for the language, it is still necessary for the rules to be put down in writing so that everyone agrees on which rules are valid. In the following part of this section, we present semantics, then a grammar, and finally a number of viewpoints on the visual detectability of symbolic languages which can be used on VDU screens.

3.4.3   SEMANTICS

Semantics specify what the different characters mean. In this case, the semantics are summarised in Figure 3.12. The notion of codes is used here as an overall concept

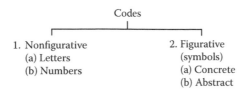

FIGURE 3.12   Categorisation of codes.

covering all types of symbols/characters. We are conventionally accustomed to two types of codes:

1. Nonfigurative codes, which consist of several small elements. The individual elements have no meaning in themselves, but when combined they have common unambiguous meanings. The most common form of these nonfigurative codes is numbers and letters.
2. There are also figurative codes (symbols), which are those designed in such a way as to have a meaning themselves without needing to be combined with other symbols. Such figurative codes may be either concrete or abstract. The concrete codes attempt to imitate what they symbolise (e.g., a pedestrian crossing sign is represented by a stylised drawing of a walking person), while the abstract codes symbolise an abstract concept (e.g., 'Christianity', represented by a cross).

Both figurative and nonfigurative codes are used in process industry applications. The figurative concrete codes should try to resemble the apparatus and machines they represent, and the abstract codes should be used to represent the actual events occurring during the process. Nonfigurative codes, usually in the form of letter and number abbreviations, are also used. Numbers are used both for identification and quantification although it is preferable to use two different number series for these purposes, for example, Roman numerals for identification and Arabic (ordinary) numbers for quantification.

3.4.4   GRAMMAR

The grammar to be used in this connection must be very simple. It should consist only of:

* Nouns
* Adjectives
* Verbs

Nouns should be used to specify different physical objects such as generators, motors, transformers, and switches. The verb is used to give the condition of the noun—for example, on, off, running, open, closed. The adjective is used either for specifying which machine/unit is under discussion, or for giving its characteristics, for example, DC, AC, size. Concrete symbols are suitable for nouns and abstract symbols are best used for verbs (arrows, colours, and so on). Nonfigurative codes such as abbreviations or numbers are best used for adjectives. Figure 3.13 shows the symbol for a lathe chuck. It has a concrete symbol to specify the noun, arrows are used to denote whether the chuck is opening or closing, and numbers and letters specify which lathe the chuck belongs to.

3.4.5   COMPREHENSIBILITY OF CODES

In this section, we look at the ease of understanding visual codes. It is, however, important to remember that sound codes can be just as useful in many instances as visual codes. One advantage of sound codes is that it is not necessary to be in a particular workplace in order to notice them, and also faster reactions may be expected to sound signals. Hearing can be said to be the dominant sense in terms of noticing, recognising, and identifying patterns. The ability of a musician to recognise very small variations in a piece of music (which is a pattern presented in the form of sound) is outstanding. The amount of information and the total number of possible different messages that could be presented in this way is immense. Sound also has the advantage in that it is not necessary to focus attention on the equipment supplying the information material, but one is free to move around while listening for the information.

Another information channel that gives almost the same degree of freedom is the sense of touch. There have been experiments to produce a language for the deaf, which is transmitted through touch. There has been an attempt (see Ivergård,

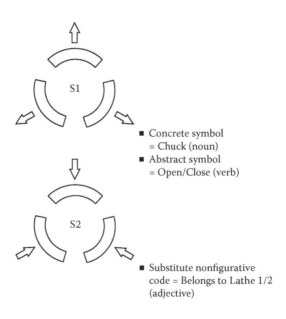

- Concrete symbol
  = Chuck (noun)
- Abstract symbol
  = Open/Close (verb)

- Substitute nonfigurative
  code = Belongs to Lathe 1/2
  (adjective)

FIGURE 3.13   Symbol for a lathe chuck.

1982, for example) to produce a 'hearing glove'; this has an instrument mounted against each finger, which produces a pressure at different frequencies on the fingers. Sound—for example, from someone talking—is converted electronically and transferred to this hearing glove. It is known that at least four different frequencies can be sensed, so with this method one can transfer at least 4 x 4 different symbols (information units). Similar methods have also been used by placing the vibration devices on the chest and other parts of the body.

The sense of touch is not only useful as an information transmission channel for the deaf but it can also be used for people who are overloaded with visual or sound information. Pilots, for example, are often overloaded by the large number of instruments they have to read. If we wish to give the pilot even more information, vibrators could be used to transfer information through the sense of touch instead. Like hearing, touch is better than sight for use as a warning signal. Whereas one can close the eyes to remove the sense of vision, the sense of hearing or feeling cannot be shut off. Four variables can be used in the transferral of information by touch: the positioning, frequency, amplitude, and variation of the stimulation.

The visibility of visual codes will not be discussed in detail here except to say that they must be (a) large enough, (b) bright enough, and (c) have sufficient contrast, in order to be visible.

Experience from Gestalt psychology can be useful regarding the suitability of codes and for general rules on their design. In addition, there are a number of perceptual psychological grounds that form the basis of the number of elements which can be used in different codes in order to avoid confusion. In particular, there is a body of specialist knowledge on the design of letters and numbers.

Table 3.3 summarises the number of possible variants that can be obtained with different types of stimuli for designing codes. It is important to remember that the estimations given are only approximations. The maximum number only refers to conditions where the person performing the reading has special education or training for that particular form of stimulus. In most normal applications—for example, work on VDUs—the recommended number of alternatives must be used. Table 3.3 also gives the number of absolutely recognisable units of a particular stimulus. If comparisons are possible, the number of recognisable units becomes considerably greater.

Table 3.4 identifies letters and numbers that are easily confused with each other, including also those that are difficult to read. The risk of confusion on VDU screens is especially great, as the letters and numbers are built up in a fairly simple and limited manner. Figure 3.14 shows the number of seconds it takes to read a particular number compared with the grouping of the digits. It is clear from Figure 3.14 that groups of three digits are the quickest to read. Suggested forms of grouping are given in Table 3.5, which gives examples of how signs in codes of different lengths should be grouped.

3.5   USING COLOUR

First, it is necessary to define what is meant by colour, as various colour classification methods exist. Probably the best-known colour classification system is the CIE

TABLE 3.3
Number of Recognisable Units of Different Stimuli

Stimuli	Maximum	Recommended	Notes
Colours:			
Lamps	10	3	Good
Surfaces	50	9	Good
Design:			
Letters and numbers	∞	?	
Geometrical	15	5	
Figures/diagrams	30	10	Good
Size:			
Surfaces	6	3	Satisfactory
Length	6	3	Satisfactory
Lightness	4	2	Poor
Frequency (flashing)	4	2	Poor
Slope:			
Direction of pointer on dial	24	12	Good
Sound:			
Frequency	(Large)	5	Good
Loudness	(Large)	2	Satisfactory

```
              1            2            3
Violet      Blue        Green      Yellow    Red
 ├──┬──┬──┬──┬──┬──┬──┬──┬──┬──┬──┬──┬──┤
 1     2     3     4     5     6  7     8        9
```

system.* This has many advantages when used for specifying colours for use on VDU screens. This system is described in Wyszecki and Stiles (1967).

The colours on a CRT are created using three electronic 'guns' that are aimed at the front of the CRT.† This is covered with a layer of phosphor. Each gun produces a different colour—red, green, or blue. The rays travel through a shadow mask that has many small holes in it, and then meet the front of the screen where there are a large number of symmetrically arranged round phosphor dots (see Figure 3.15). This results in different combinations of the colours red, blue, and green. Because the points are very close together, the human eye does not perceive them as individual points but as mixtures of colour.

Rather than a spatial combination of colours, one can have a temporal combination. This means presenting the different colours at different times. Because the succession time of the colours is very short, the eye will not be able to pick out

---

* Commission Internationale de l'Eclairage (International Commission on Illumination).
† Other types of visual display technologies use a similar logic to produce different colours on a screen.

TABLE 3.4
Interchanging of Letters and Numbers

These: are mistaken:	O (zero)	8″	B′	D′	I″	Ø″ capital O	Z″	S	G	N	V	Y
for these:	Ø	B	0 (zero)	0 (zero)	1	0 (zero)	2	5	6	W	U	X
	6		Q	D		D		3	C		Y	V
	D		P									4
	9											T

′Interchanged often; ″Interchanged very often.
The following are difficult to read: 2, 4, 5, 8, 9, N, T, I, Q, X, and K.

On VDU Screens

Mutual		One-Way	
O and Q		C	read as G
T and Y		D	read as B
S and 5		H	read as M or N
I and L		J, T	read as I
X and K		K	read as R
I* and 1*		2	read as Z*
		B	read as R, S, or 8*

*These four account for more than 50% of all errors.

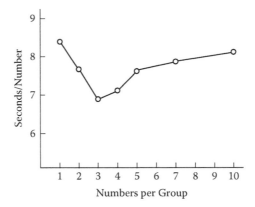

FIGURE 3.14   Average reading time per numbers of the keying in of 18 to 21 numbers at a time as a function of their grouping.

TABLE 3.5
Coding Groups of Different Lengths

Number of Digits	Alternative Groupings		
7	3, 3, 2	3, 4	3, 2, 2
8	3, 3, 2	2, 2, 2, 2	
9	3, 3, 3		

individual colours, but they will be combined in the proportions in which they were presented. One drawback of this type of colour presentation is that it gives considerably less well-defined colours which are difficult to identify.

In the CIE system, a colour mixture will always lie on a line between the two colours of which it is a mixture. If the mixture is made up of three colours, the perceived colour will lie within a triangle formed by the lines between pairs of colours from the three. This ability to define colours on a CRT fairly simply using the CIE system is its major advantage over other descriptive classification systems. However, an obvious disadvantage of the CIE system is that colours cannot be classified in comprehensive descriptive terms such as colour, saturation, brightness, and so forth. Other descriptive classification systems do have this advantage.

There are a large number of factors that affect the perceived colour. The following are the factors that must be taken into account:

1. Contrast
2. Size
3. Background colour
4. Absolute relative discrimination
5. Number of colours
6. Surrounding lighting

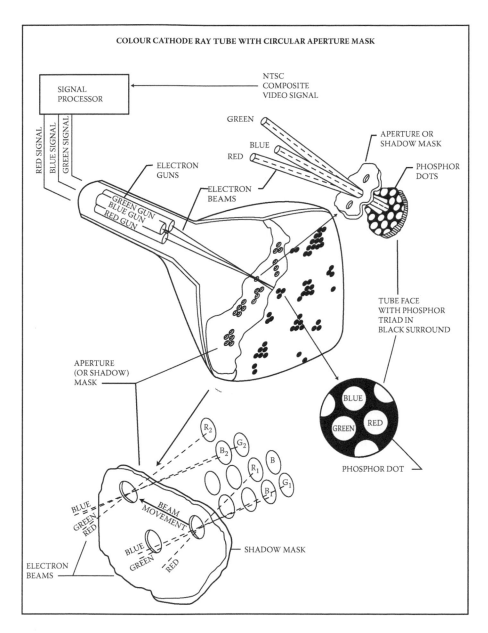

FIGURE 3.15 Colours on a CRT are created by the electronic 'guns' shown in this figure. (Reproduced from Farrell and Booth, 1975. With permission.)

When working only with black and white, it is sufficient to use luminance contrast (CR), which is normally defined as:

$$CR = L_1 / L_2 \qquad\qquad (3.3)$$

where L_1 is the luminance of the object and L_2 is the luminance of the surroundings. When colour also has to be taken into account, the concept of colour contrast is used. A special discrimination index has been worked out, which takes into account both luminance and colour contrast (see Silverstein and Mayfield, 1981).

Sensing of colour is highly dependent on the size and brightness of detail within the picture. Very small details appear less saturated and sometimes change depending on background colour. Also, the ability to discriminate between colours is severely reduced for small pictures, especially along the blue-yellow axis. The background colour is very important as well because the light from a small object is experienced as though it changes colour towards its complementary colour in the surroundings.

As the number of colours increases, colour discrimination becomes more difficult, and greater demand is made for colour coding. Research on the use of colours on VDU screens has shown that a maximum of four to six colours can be used. The number of colours which can be used is, of course, highly dependent on whether the discrimination is absolute or relative. In absolute discrimination, the number of colours is further reduced. When colour is used for decorative purposes, more colours can be used. The reasons for using colours to any great extent must be powerful ones; it is far easier to produce a poor colour picture than a poor black-and-white one.

In recent years, comprehensive research has been carried out into the use of colour screens, and the level of knowledge has increased considerably. Unfortunately some important knowledge is still missing, which makes it difficult to make final recommendations. In particular, the effects of interacting factors are not well understood. For those who wish to learn more about the analytical method used in determining colours on a colour screen, Silverstein's (1982) *Human Factors for Colour Display Systems* is recommended.

Why should a colour display be chosen? In most cases, colour displays are just a more costly alternative to the monochrome version. When there is an advantage in using a colour display, it does not seem to have an economic basis. It is probable that a colour display is *not* preferable to a monochrome display for most types of common office applications. Where there are special office applications needing more complex pictures, or in special applications such as control rooms in the process industries and traffic monitoring situations, colour displays are certainly preferable.

Selection of a colour display does not depend only on the fact that the operator prefers to have a colour screen for aesthetic reasons. The primary reason for choosing a colour display is for increased coding potential and the reduction in search time it brings about in complex pictures. It is also important to remember, as demonstrated later, that colour contrast increases visibility, and thereby decreases the need for luminance contrast. This, too, has many advantages, as colour screens can be utilised in rooms with an ordinary level of lighting which fulfils the requirements for other types of visual tasks. Under special visual conditions—for example, on planes and ships—the colour screen is almost obligatory, although the requirement for well-shielded lights still remains.

Colour VDUs drastically increase the requirement for specialist knowledge in the design of the display system and the choice of coding for pictures. It is difficult to make a good picture on a colour display. A colour picture can easily be made worse than a black-and-white picture if one does not have access to the right knowledge. A poorly designed colour display decreases efficiency of performance as compared with a black-and-white screen.

3.5.1 Choice of Colours

A number of general ergonomic rules pertaining to absolute discrimination indicate that for recognising and naming a colour, a maximum of about seven colours is desirable. When it comes to seeing the difference between colours, many different colours and shades can be distinguished, probably well over twenty. Research on colour screens has shown, however, that this is unrealistic. One cannot normally have more than three to seven colours on a CRT, depending on a number of different factors (Kinney, 1979; Teichner, 1979; Silverstein and Maryfield, 1981). For an operational colour display where absolute colour discrimination is required, it is suitable to have three to four colours, and these should preferably be green, red, blue-green, and possibly a purple-red (plus white or black, depending on the background colour used). Where comparison between colours is possible, six to seven colours can be allowed, preferably red, yellow, green, blue-green (cyan), reddish-blue, and perhaps magenta (plus white or black depending on the background).

3.5.2 Colour Screen Character and Symbol Design

It is fairly clear that the requirements will be far more stringent for the design of characters and symbols on a colour screen than for those on a monochrome display with a dark background and light text. Research has shown that colour sensitivity increases as the colour field increases in angle, up to 10 minutes of arc. Small light fields have a reduced colour saturation which means that in practice light colours tend to be seen as white.

On this basis, characters, symbols, or critical details in the picture that subtend less than 20 minutes of arc should not be used. On larger fields, where greater accuracy in colour discrimination is required, this angle should be at least 1°. The lines that make up the characters should be thicker on a colour screen than on a conventional one. If there is more than 1 minute of arc between lines, colour separation will be distinguishable. The characters on a colour screen should not be placed more than 15 minutes of arc from the line of sight. Characters that appear towards the periphery of the screen will not be colour-coded to the same accuracy.

Research has also been performed on legibility and luminance contrast of colour screens under different environmental lighting conditions (Silverstein, 1982). This has shown that there is no further improvement in legibility when the luminance contrast is increased beyond 5. It should also be remembered that the colours seen on the screen will change depending on the room lighting. The room lighting should therefore not be changed once the colour screens are installed.

3.5.3   VDU SCREEN AND BACKGROUND REQUIREMENTS

The technical requirements for a colour screen are considerably greater than for an ordinary monochrome VDU screen. One must, for example, set considerably greater demands on sharpness of edges and the ability of the 'guns' to send their rays to the correct place at the edges and corners of the screen, so that no convergence problems occur. It has been shown that increased luminance on a VDU screen increases the colour contrast up to a luminance on the screen of 3000 cd/m^2. Increasing the operator's light adaptation level also increases his or her sensitivity to colour (Silverstein, 1982). A light background gives the effect of a higher degree of colour saturation. In practice this means that a higher level of environmental light can be used when working with colour screens. In many cases it may also be desirable to select a light background colour on the VDU screen in order to achieve an even higher level of saturation which, in turn, reduces the risk of both diffused and reflected glare.

3.6   SPEECH RECOGNITION AND SPEECH GENERATION

Language is probably the most distinguishable ability that mankind possesses. Human/machine systems have long included written language (for example, in alphanumeric displays and keyboards), but spoken language has mainly been used for communication between people. Over the past few decades, automatic speech generation and recognition by machine now offers a possible alternative to other forms of input and output to computers. However, there are a number of limitations that need to be overcome. Knowledge about how to use speech recognition is developing very fast; in some special areas, it can be a very good alternative, particularly in control room situations.

An interactive speech system consists of speech recognition devices for control or information input and speech generation devices as a form of information display.

Automatic speech technology is of great interest already today, but there are many problems that need to be resolved. In 1985, Simpson et al. gave a comprehensive review of *system design* for speech recognition and generation. From the human factor/ergonomic perspective, a speech recognition system consists of a human speaker, recognition algorithms, and a device that responds appropriately to the recognised speech:

Human speech → Recognition algorithms → Response device

Today, many different types of algorithms can be used to interpret human speech. An important feature of many of these algorithms is that they can *learn* to recognise the sounds of different voices and accents. In ideal conditions many algorithms can recognise more than 90% of speech. When situations are less than ideal—for example, in a noisy environment or when a speaker has an unusual dialect—the recognition might be severely impeded. In a control room situation, one can expect the use of a rather limited vocabulary, for example, in emergency situations. With a limited vocabulary the speech algorithm can be developed to recognise many different accents, even when the background environment is noisy and disruptive.

A speech generation system is the mirror image of a recognition system. It consists of a device to generate messages in the form of symbol strings, a speech generation algorithm to convert the symbol strings to an acoustic imitation of human speech, and a human listener. A speech generation system operates within the context of the user's working environment:

Message generator → Transformation of message codes to acoustic
imitation voice → Human listener

Many factors contribute to the operational intelligibility of speech. Simpson et al. (1985) proposed a model for operation intelligibility (Figure 3.16). It is important to note that intelligibility and the characteristics of human speech are not necessarily correlated. A radio announcer may sound natural despite a background of static noise but may have low intelligibility for the listener. Conversely, synthesised speech warning messages in an aircraft cockpit may sound mechanical, but pilots consider them to be more intelligible than 'live' messages received over the aircraft radio.

Comprehensive design guidelines in this area are still very limited, but some general points can be made. Although the ergonomics/human factors literature includes reports of research, which supports certain principles of speech design, this knowledge has not yet been formulated into design guidelines. Human factors methodology is sufficiently well developed to permit comparison of task-specific speech systems experimentally, but the tools required for producing generic design guidelines for speech systems are not yet available. In the short term, simulation of speech system capabilities in conjunction with the development of improved system performance should prove a productive methodology to achieve these aims.

In general, speech generation algorithms seem to be more advanced than speech recognition algorithms. Reasonably intelligible text-to-speech algorithms from

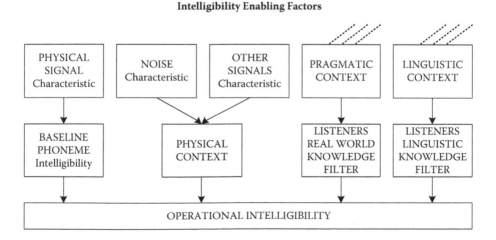

FIGURE 3.16 Factors that contribute to operational intelligibility. (Simpson et al., 1985. Redrawn from *Human Factors*, 23, p. 131. © The Human Factors Society, Inc. With permission.)

Standard English spelling are now available commercially. The recognition counterpart, speech-to-text algorithms, is also available in the form of dictation systems. These different available systems seem to be heavily user-dependent. Some people regard the quality as very high while for others it is unacceptably low. Systems transferring speech into control actions have a higher reliability, particularly if the expected vocabulary is limited. For example, in an emergency, when the operator shouts 'Stop generator!' the system will recognise the instruction and act accordingly.

In the short-term, the current recognition algorithms appear adequate for use in favourable environments characterised by low to moderate noise levels, for applications that only require small vocabularies and that do not place the operator under stress. Great caution must be exercised with the use of current technology in stressful situations.

Speech generation algorithms, on the other hand, have demonstrated acceptable performance even under conditions of severe noise and high workload. This technology is sufficiently advanced to be applied appropriately, with careful attention being paid to the integration of ergonomics. Already by 1980, Simpson and Williams had studied the use of synthesised voice for cockpit warnings. They concluded that voice warnings do not need to be preceded by an alerting tone, as the quality of the spoken language makes the warning readily understandable by the listener.

REFERENCES AND FURTHER READING

Berns, T., and Herring, V. (1985). *Positive versus Negative Image Polarity of Visual Display Screens*, Ergolab report 85:06. Stockholm: Ergolab.

Cakir, A., Hart, D., and Stewart, T. (1979). *The VDT Manual*. Darmstadt: IFRA.

Conrad, R., and Hull, A. (1968). The Preferred Layout for Numeral Data-entry Keysets. *Ergonomics* 11, 165–73.

Easterby, R.S. (1965). *Evaluation of the British Standard Proposals for Symbols on Machine Indicator Plates*. Cranfield: College of Aeronautics.

Easterby, R. (1967). Perceptual Organization in Static Displays for Man/Machine Systems. *Ergonomics* 10, 195–205.

Easterby, R. (1968). *The Perception of Symbol Displays*, AP Note 13. Birmingham: Aston University.

Farrell, R.J., and Booth, J.M. (1975). *Design Handbook for Imagery Interpretation Equipment*. Seattle: Boeing Aerospace.

Granit, R. (1965). *Receptors and Sensory Perception*. New Haven, CT: Yale University Press.

Ivergård, T., Istance, H., and Günther, C. (1980). Datorisering inom Process Industrin, Part 1. Stockholm: ERGOLAB.

Ivergård, Toni. (1982). *Information Ergonomics*. Lund, Sweden: Studentlitteratur.

Kaimann, Richard A. (1993). Ergonomically Designed Keyboards: Ready or Not, Here They Come! *Journal of Systems Management* 44, 4 (April), 16–27.

Karlsson, B., Karlsson, N., and Wide, P. (2000). A Dynamic Safety System Based on Sensor Fusion. *Journal of Intelligent Manufacturing* 11, 5 (October), 475–83.

Kinney, J.S. (1979). The Use of Color in Wide-angle Displays. *Proceedings of the Society for Information Display* 20, 33–40.

Murrell, K.F.H. (1958). *Fitting the Job to the Worker: A Study of American and European Research into Working Conditions in Industrial Situations*. Paris: Organisation for European Economic Co-operation–EPA.

Seminara, J.L. (1980). *Human Factors Considerations for Advanced Control Board Design.* Vol. 4, *Human Factors Methods for Nuclear Control Room Design.* Palo Alto, CA: Electric Power Research Institute.

Shaikevich, A.S. (1959). Classification of Visual Tasks. *Svetatecknika* 13, 91–95.

Silverstein, L. (1982). *Human Factors for Color CRT Displays.* San Diego, CA: Society for Information Display.

Silverstein, L. (1987). Human Factors for Color Display Systems. In H.J. Durrett (ed.), *Color and the Computer,* 27–42. Orlando, FL: Academic Press.

Silverstein, L.D., and Maryfield, R.M. (1981). *Color Selection and Verification Testing for Airborne Color CRT Displays.* Proceedings of the 5th Advanced Aircrew Display Symposium, September, Naval Air Test Center.

Simpson, C.A., McCouley, M.E., Poland, E.F., Ruth, J.C., and Williges, B.H. (1985). Systems Design for Speech Recognition and Generation. *Human Factors* 27, 115–41.

Simpson, C.A., and Navarro, T.N. (1984). Intelligibility of Computer Generated Speech as a Function of Multiple Factors. In *Proceedings of National Aerospace and Electronics Conference,* 932–40. New York: IEEE.

Simpson, C.A., and Williams, D. (1980). Response Time Effects of Alerting Tone and Semantic Context for Synthesized Voice Cockpit Warnings. *Human Factors* 22, 319–30.

Singleton, W.T. (1967). Systems Prototype and Its Design Problems. *Ergonomics* 10, 120–24.

Smith, K.D., and Walker, B.A. (2001). Optimum Console Design Promotes Control Room Efficiency. *Hydrocarbon Processing* (September), 163–70.

Stark, Lawrence. (1983). Personal communication with Toni Ivergård.

Stewart, T.F.M. (1976). Displays and the Software Interface. *Applied Ergonomics* 9, 137–46.

Teichner, W.H. (1979). Color and Information Coding. *Proceedings of the Society for Information Display* 20, 3–9.

Wyszecki, G., and Stiles, W.S. (1967). *Color Science: Concepts and Methods, Quantitative Data and Formulas.* New York: John Wiley & Sons.

4 Design of Large and Complex Display Systems

Eric Hénique, Soeren Lindegaard, and Brian Hunt

CONTENTS

4.1   INTRODUCTION

Control rooms and control centres tend to have something in common, even though control rooms are used in several areas of application. The specific types of applications are many and varied and include telecommunication, military and aerospace, industry, energy (generation, distribution, and transmission), security (national and domestic), water (production, sewage, and purification), traffic monitoring and control (roads, tunnels), airports, railway and metro networks, police and fire departments, emergency services and call centres, data centres for governments, and commercial enterprises. In general, the central medium of such a room nowadays is a large screen system. Other terms for large screen display include: video wall, large display screen, large video screen, display wall, large video screen, large screen display, and monitor wall. Other technical and specific terms used in connection with this display screen technology are contained in a glossary at the end of this chapter.

The increasing volume of traffic and the resulting traffic jams, the growing traffic network, and environmental aspects make observation and control of the traffic-related systems necessary. Data and information transmitted from traffic detectors, video observation systems, and other sources have to be administered, distributed, and visualised centrally in traffic management centres. This guarantees a timely control and administration of this data and thus a productive and efficient use of the existing traffic resources. In recent years, headline-catching accidents and fires in road and train tunnels have encouraged governments to launch new programmes to reinforce the security in tunnels. As a result of such initiatives, significant investments in control room systems have been made and are still going on. This makes it also necessary to invest in key control room equipment such as large screen systems, which can help reduce dramatically the reaction times to problems.

Effective control and security of transportation systems depends substantially on the availability of all necessary information in the control rooms. Safety and the smooth operations of roads, highways, railways, and air traffic are facilitated by competent operators manning the control rooms. In railway networks, the stations, platforms and station concourses, tram termini and crossings, and subway networks warrant close attention from the control room operators. On waterways, the bridges, canals, and shipping flow are key areas where vigilance ensures efficient operations. Air traffic controllers ensure safety in the air while their colleagues monitor pedestrian and vehicular traffic around and within the terminal approach roads and buildings. A key resource for these professionals in their work is the large screen wall that is the central unit for all operational tasks in the control room. All applications, software, and video signals can be displayed in real time on the projection wall, so they can be rapidly processed and transmitted by the operators.

Large screen walls are used in these operational fields for the following applications: graphics (maps, road and railway networks, and time schedules), GPS applications, camera signals (traffic detectors, signal systems on main traffic junctions, railways, and airports), observation systems, alarm management, monitoring of the ventilation systems in tunnels, remote control, and other transport systems and software.

In the crucial activities involved in water supply, purification, and production, a growing demand for water, energy, sewage facilities, and production capacities has led to a growing need for increasing amounts of relevant data and information. Quantities of data have increased hand-in-hand with rapid changes in information technology and processing. Nowadays, the most common systems used in management and control centres are SCADA (supervisory control and data acquisition) systems and DCS (distributed control system) in combination with large screen displays. Using these technologies, the operators can observe in real-time flow diagrams, pumps, valves, pressures, reservoirs, metres, status information, and other relevant data. At the same time, observation cameras situated at strategic points in the processes make it possible to visualise the state of the process and relay signals to control the process flows.

Since this enables a total surveillance of the entire production facility at any time, it is possible to identify failures and to react on malfunctions with very short response times. Given the strategic nature of this industry, the security aspect is in the foreground. This factor has an important influence on the productivity and efficiency of a plant. Similar situations apply in the observation and control of nuclear power plants, water plants, gas and oil distribution facilities, production plants, as well as waste services such as sewerage plants. In these industrial processes, security and safety are tightly connected with the image quality and functionality of the master display in the control room.

The most common applications that are displayed, edited, and managed on large screen displays are SCADA or DCS applications (pipelines, production networks, pump stations, and so forth), graphics (production survey, networks, customer-specific applications, and so on), observation cameras, and further supplying and distributing applications.

Large screen walls also play a key role in the control rooms and the data centres of the telecommunication (network operating centres [NOCs]) branch of fixed or mobile phone companies, and sometimes even the world-spanning networks. Complex graphics (maps, network diagrams, and customer-specific applications), high-quality voice and data streams, network resources, fault management (network control, service, fault removal, availability), security management (network security, redundancy, tests), accounting management (exact account calculation, account surveillance), and quality management are controlled and observed on the large screen displays in control rooms of the telecommunications industries.

Rising levels of criminal and terrorist activities, not only since the incidents of September 11th, have predictably led to a growing demand of observation in the security sector. In security centres the relevant data and video signals have to be available for continuous (24 hours a day, 7 days per week) operation in real time. Most important for a reliable security system is a perfect interaction of observation, control, management, planning, and coordination of activities. Therefore the concerned services all operate interconnected to control and coordinate their activities. Some recent terrorist attempts had been averted because of the closed-circuit television (CCTV) surveillance. Police officers could quickly apprehend suspected terrorists because of this surveillance and the alertness of the control room operators. In the military sector, surveillance can be conducted via surveillance input provided by unmanned aerial vehicles (UAVs), also called drones. Cameras mounted on the drones are able to record real-time images from high or low altitudes. Different types of lenses fitted to the cameras allow the images to be taken by day, at night, and through cloud cover. Relayed by satellite to large screens in command and control centres and collated with incoming data from other sources, the images enable relevant decision makers to deploy available personnel and weaponry in the air and on the ground.

Apart from these roles, large display systems are used to improve the reaction time and to increase the output in data centres and production facilities. The use of large display systems has played a role in constantly increasing productivity and efficiency. Again, the central medium in all those control rooms is a large screen system. Such a large screen system is in general used to collect, visualise, and distribute data and information to give a complete overview of the ongoing situations to a number of users or operators. The management is also quickly able to receive an overview on what is happening inside the control room.

4.2   APPLICATIONS FOR LARGE SCREEN SYSTEMS IN CONTROL ROOMS

Companies in various industry sectors are using large screen systems in their control rooms, where the display of process data in real time is of critical importance. Thus this technology is used by a range of process industries such as energy exploration, production, distribution, and waste disposal (for example, sewerage works, waterworks, power plants, oil and gas transportation, public utilities, and waste incineration). Transportation industries that also have process flows (of vehicles and people rather than minerals or liquids) also use large screen displays. Transport networks (including roads, highways, metro systems, railways, tramways, cross-channel ferry services, tunnels, airports, waterways, bridges, and traffic management centres) use this technology to monitor and control optimum flows in their systems.

Apart from process monitoring and control, other applications in industry and industrial processes include company security, for example, in security-conscious areas such as core refinery areas and manufacturing sites for nuclear fuels. Large screens also play a key role in workstations for the design and development of collaborative projects, conferences (including teleconference facilities), and company presentations (both in-house and virtual).

An increasing area of use of this technology is in the security and surveillance industry, in both the government and private sectors. There is increasing use of CCTV surveillance in public environments such as town centres, shopping malls, roads and motorways, and car parks as well as semiprivate spaces such as offices, sports centres, lifts, and government buildings. In spite of protests from organisations concerned with civil liberties, the use of CCTV is predicted to grow. Private security companies that offer organisations security and protection services have long used large screens in the control rooms for their operations. The control rooms of security centres (for example, in shopping malls) and of the emergency services (police, the fire service, and hospital and ambulance services) all use large screen displays in their central coordinating facilities.

The telecommunications industry has been a long-term user of large screens in its operations. These include: network operations centres, video conferencing, the control and management of data service centres, and network observation). Other areas for large screen use include simulations and the various industries involved in broadcasting and entertainment.

4.3   WHY USE LARGE SCREENS IN CONTROL ROOMS?

The main reasons for using a large screen system inside a control room include ergonomic aspects to provide a better work environment for the control room operators by the use of equipment to enable people to be more comfortable and have a better overview of all applications and processes. The display of data and information in a large format means that there can be a reduction of reaction times (for example, in teamwork where all members can see the same information on screen). For the organisation, there are likely to be issues of savings in operating costs, and the perceived need to increase employee efficiency and productivity through sharing information and collaborative work. Included in the latter is the possibility for multiple displays, giving the operators access to increased levels of surveillance and monitoring. Combining additional functions and multiple functions could lead to intelligent displays as data and information are combined for greater utility. Operators could benefit from receiving a high quantity of data and information from different sources that they could then combine and use to improve their decision making.

4.3.1 DEFINING REQUIREMENTS

For the use of a large screen system in a control room, different factors have to be considered. These factors include the physical location of the large screen in the control room, with consideration for the available space and the nature and duration of visualisation and operation. Different considerations will prevail if the screen is to be used on a full-time (24/7) operational basis or part-time (for example, only during certain work shifts or usage patterns). Another consideration is the type and quality of displayed sources and, in particular, whether these will be from video or computer sources. In deciding the location of the screen, consideration needs to be given to the physical conditions in the control room, such as the lighting conditions (if the light source is artificial or natural) and the position of windows and skylights providing any natural lighting. Risk of glare is very important, as glare (for example, daylight reflection) could completely kill the visibility of the screen. Particular attention needs to be given to elements of the room construction, such as the weight-bearing capability of the floor and the construction of the walls. The technology of large screen displays requires certain environmental conditions, including a dust-free environment and a robust base construction that keeps the screen stable. Given that space is a critical issue, thought should be given to the position and number of access points for the control room operators. The number of operators and users at any one time is also an influencing factor, as this affects sight lines to the screen and the optimal location of other equipment that the operators may need. Overcrowding is to be avoided. It is also important to have a clear allocation of responsibility between the operators to prevent confusion and work duplication.

In terms of defining needs, there are many features of the large screen display that should be considered while planning the use of the technology and certainly before purchasing the equipment. The size of the large display screen and the space needed for its effective use should not be underestimated. The height of the room is also critical and needs to be factored into calculations, as does the height of surrounding furniture to be placed in the room. Consideration should be given to the distance that the operators are expected to be from the screen and whether their view is impeded by intervening furniture. The distance that the operators are from the display cubes that make up the large display screen will affect the operators' usability of the technology.

In terms of needs of the technology, the size and resolution of cubes in the display are a key issue in the suitability of the equipment to the size of the room in which it is located and the ease with which the users can view the displayed data. Large display screens are available in curved and plane surface formats. The available space overall and the location of the control room operators in relation to the screen display will be deciding factors in the choice of these formats. Other contributing factors are the number and type of resources to be used in conjunction with the display screen.

When purchasing a large screen, natural areas of inquiry will include budget (especially any budgetary restrictions or constraints, such as those imposed by a head office on a regional operation). Apart from deciding the budget available for the equipment and including the projected costs of maintenance, it is wise to make a site visit to make a detailed inspection of the equipment in use. In this way the potential purchaser can see the actual dimensions of the equipment in the workplace. Included in budgetary issues are projected service and maintenance costs. Technical questions include the durability of the screen, the required resolution of the displayed sources, and the visibility and readability of the data and information displayed on the screen (given the location factors and the size of the room and number of users). Given Moore's law* that technology improves exponentially every two years, it would be prudent to check the equipment for expandability for a later upgrade.

4.3.2 VISUALISED INFORMATION

As most of these control rooms are used for continuous operation, ergonomic aspects are also very important. This includes the design and usability of the lighting and furniture (consoles, chairs, and work surfaces), as well as the technological aspects (with a 24/7 capability). Depending on the technology, the following information and data can be displayed on a large display system: SCADA and DCS systems, maps, grids, networks (for example, power grids, pipelines, roads, and rail networks), video sources (CCTV cameras, Internet protocol [IP] cameras, DVD, and VCR), servers or workstations, production processes, presentations, timetables, surveillance systems, alarms, flow diagrams, and status messages.

In principle all kinds of electronic information should be possible to be displayed. The different areas (for example, telecom, energy, security, and so forth) have different requirements, but also within the same area the requirements cannot be generalised. Every control room project is unique and has its own specific requirements that have to be considered in the design.

4.4 BASICS OF VIEWING AND SEEING

Before exploring the different technologies, it is helpful to describe some basics of viewing and seeing (see also Chapter 10 of this handbook).

4.4.1 THE HUMAN EYE

The human eye has two sets of photoreceptors, the rods and the cones. The rods are active at very low light levels and the cones at high light levels. Perception is the process in the human brain that interprets the impulses from the photoreceptors into images that can be processed and understood within the mind. Human vision is the primary of our five senses, handling more than 90% of the inputs to our nervous system. Apparently, the sense of vision tends to be overloaded and all other sense organs are not used to their fullest capacity. In the future, we must look into the possibility for using our other sense organs (such as touch and hearing).

---

* In 1965, Gordon Moore, then R&D director at Fairchild Semiconductor and later cofounder of silicon chip manufacturer Intel Corporation, stated that the density of transistors that can feasibly be placed on an integrated circuit chip doubles every 18 months. This 'law' was later refined to suggest that technology increases in performance every 2 years (see, for example, Grove, 1990; Schaller, 1997).

Human vision is a highly-sophisticated and integrated system, capable of handling an enormous amount of information, adapting to extreme light conditions, and distinguishing between minute details. For this reason the quality of the displays is crucial. The human eye has a tremendous, albeit limited, capability to adapt to extreme variations in light level ranging from star- or moonlight to direct sunlight. There is also potential to use other sense organs. Muscular receptors and sense organs have a very large potential, particularly for the development of tacit abilities. However, this will also demand the use of other types of controls than keyboards (Figure 4.1).

While the rods provide achromatic (noncolour) vision in the range of illumination levels from 10e-6 to 10 cd/m², the cones provide chromatic (colour) vision in the range 0.01 to 10e8 cd/m². The human eye is capable of detecting illumination levels with a difference of log 14, or 100,000,000,000,000 times. However, the neural units processing the signals to the brain are only capable of transmitting a signal with a dynamic range of log 1.5. When the eye adapts to the average light level present, it is to some extent dynamic. But when the light level changes too fast, the eye becomes momentarily blinded. A typical example of a sudden change in light level is turning on the light in a completely dark room. This sudden change in environment affects the eye's adaptation time, and thus results in temporary blinding.

How the eye adapts to the given light level is based on the average light level in the field of vision. This *average adaptation level* and all light sources are related to this adaptation level. At night the headlights of an oncoming car blind us, whereas during daytime the same phenomena have little effect on our vision. The brightness of the headlights remains the same but during daylight the adaptation level of our eyesight is much higher than during night-time. The adaptation level is a very crucial factor when designing bright, high-quality displays. The human eye perceives different luminance levels in a logarithmic or relative way, which means that a clearly perceivable difference in luminance equals a doubling of the light. A light source with a luminance of 100 cd/m² will be perceived as clearly brighter that the one with a luminance of 50 cd/m². To ensure the same perceived difference in brightness requires a step upwards to a luminance of 200 cd/m². This manner of adaptation is

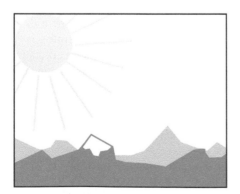 

FIGURE 4.1   Variations in the seeing ability of the human eye. **(See colour insert following page 110.)**

the reason why the eye is capable of handling such extreme differences (that is, contrasts) in light levels (Figure 4.2 and Figure 4.3).

The logarithmic characteristic of how the human eye adapts to different light levels is very important when it comes to understanding projector performance and perceived image brightness.

4.4.2 CONTRAST

Contrast enables the human brain to process perceived images. Without contrast, images cannot be perceived by the human brain. For example, it would be impossible to perceive anything in a completely white room with totally even light on all

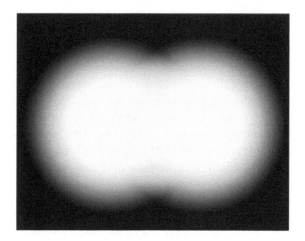

FIGURE 4.2   Perception and the human eye. **(See colour insert.)**

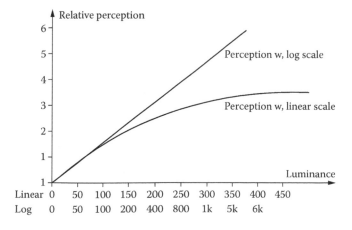

FIGURE 4.3   Relative perception and luminance.

surfaces. The eye will still see the light, nothing is invisible; it is just not possible to distinguish one thing from another. The essential ingredient for perception is contrast and, in this example, there is a lack of contrast (Figure 4.4).

The eye is capable of registering a contrast range of approximately 1000:1. This means that light sources with a luminance value below 1/30th of the adaptation level will be perceived as black, and light sources with a luminance value exceeding 30 times the adaptation level will be perceived as white. This limitation explains why we have difficulty perceiving details in the dark shadows on a bright and sunny day or distinguishing details in a bright cloudy sky.

4.4.3 COLOURS

Colours comprise a range of light at different wavelengths, and the perception of colours is dependent on both the adaptation level and contrast in the perceived image (Figure 4.5).

If the contrast level drops, the colours will be washed out. A similar effect can be seen by pouring water into a glass of red fruit juice; the dilution effect of the clear liquid (the water) changes the colour of the darker liquid (the fruit juice). Without black (that is, darkness) it is impossible to have colours. The actual balance of colours in a given light source is the *colour temperature* measured in degrees Kelvin. Pure daylight white has a colour temperature of 5600 Kelvin whereas standard TV studio lighting has a colour temperature of 3200 Kelvin. The lower the colour temperature, the more dominance there will be at the red end of the spectrum (Figure 4.6a,b).

To some extent the eye adapts to the colour temperature, which is the reason why white can be perceived as white even though the colour temperature has changed.

A typical example is the eye's adaptation to standard light bulbs, with a low colour temperature, but still the colours can be perceived almost as we would in pure white light. When comparing the two different 'whites' and perceiving the colours

FIGURE 4.4   The same image against contrasting background.

FIGURE 4.5   The colour spectrum. **(See colour insert.)**

simultaneously, the low-temperature 'white' is clearly reddish or yellow compared to the pure white.

4.4.4   VIEWING

To view displays, we look at an object and then perceive some kind of information. Using displays is a visual method of communication: the transfer of information to be perceived visually by the viewer (Figure 4.7).

Selecting the ideal display is therefore based on the requirements for visual communication. What kinds of information have to be transferred to the viewers? There is a huge difference in the requirements between a display in the airport directing passengers to the right gate and the display used in the control room, viewing such data and information as spreadsheets, graphs, production details, overviews, SCADA, and CCTV. This leads us to the question, what are the fundamentals of viewing?

4.4.4.1   Fundamentals of Viewing

The first fundamental to consider is the amount of information in a single image (Figure 4.8).

The basic requirement for how much information can be viewed at any one time is that the information should be 'perceivable'. When the information is textual, this becomes a matter of the size of the typeface in relation to the viewing distance. With more information and a smaller font used in the image, the viewing distance will need to be shorter (that is, closer to the screen).

Resolution is another way of describing the amount of information in a display. The more information needed, the higher the resolution required.

(a)

(b)

FIGURE 4.6 (a) An image in 3200 Kelvin. (b) The same image in 5600 Kelvin. **(See colour insert.)**

4.4.4.2   Viewing Angles

The next set of fundamentals includes the viewing angles (Figure 4.9). The DIN* standard states that viewing angles (that is, viewing positions) should be within ±45° and within ±30° vertically. Of course this depends also on the actual application; for example, the height of the image will depend on the seating layout and the arrangement of the work consoles.

---

* Deutsches Institut fur Normung (the German Institute for Standardisation).

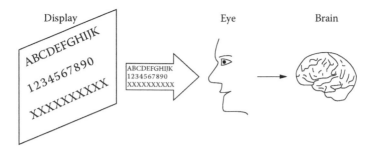

FIGURE 4.7 Visual images transferred via the eye to the brain.

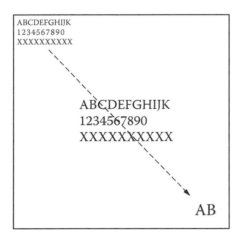

FIGURE 4.8 Perceiving text.

4.4.4.3 Brightness and Contrast

Image brightness and obtainable contrast ratios are *the* fundamentals for image quality. As can be seen from the way in which the eye works, the image brightness and contrast ratio of the screen display must be selected based on the average adaptation level in the specific application. The higher the adaptation level, the brighter the image that is required. Adaptation level is a result of a number of different factors, including ambient light level, reflection factors on the surfaces in the field of vision, and luminance of different light sources in the field of vision. Given these variables, there are no standard recommendations for the required image brightness. However, as a general rule the image brightness should be at least at the level of the average luminance level in the field of vision, provided that a sufficient contrast ratio can be obtained. For simple displays (that is, those comprising simple graphic information), the contrast ratio should be at least 5:1. High-quality graphics with saturated colours and video images require a much higher contrast ratio. As a general rule the image contrast ratio should be at least 10:1.

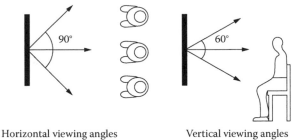

Horizontal viewing angles   Vertical viewing angles

FIGURE 4.9   Viewing angles.

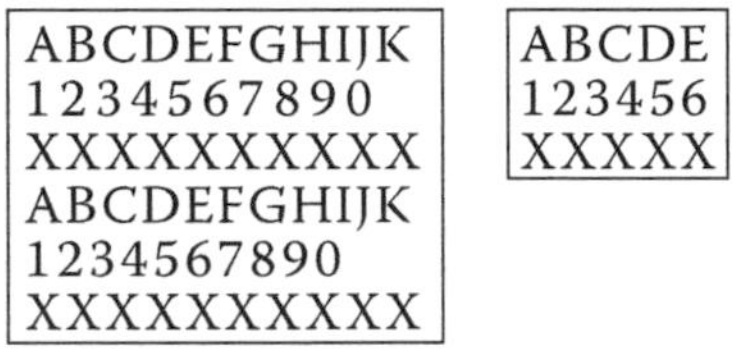

```
ABCDEFGHIJK     ABCDE
1234567890      123456
XXXXXXXXXX      XXXXX
ABCDEFGHIJK
1234567890
XXXXXXXXXX
```

FIGURE 4.10   Viewing distances.

4.4.4.4   Viewing Distances

Viewing distances are often determined by physical restrictions, such as ceiling height or the height of the consoles in the control room. Ideally the display size should be selected according to the amount of information required and the required or available viewing distance (Figure 4.10).

The more information with a given viewing distance, the larger the image that is required. If the number of viewers is high, a larger image is required. A single letter or digit might be perceivable up to a distance, which is 300 to 400 times the height of the letter; for example, if the image content is only a single letter, the screen size in relation to the viewing distance could be very small (Figure 4.11).

Studies have shown that the minimum required character size must be greater than a 16-arc minute, that is, 16/60°. This implies that the maximum viewing distance is 215 times the height of the character. What the eye is capable of perceiving is also called acuity. Therefore, a text displayed on a standard 17-inch PC cathode ray tube (CRT) monitor running at XGA resolutions (1024×768 pixels), the height of the individual letters will be approximately 2.7 mm. This will imply a maximum viewing distance of approximately 1200 mm (1.2 m). As the image height of the PC monitor is 240 mm, the ratio between viewing distance and image height is 5:1 (Figure 4.12).

Selecting the right display size should be considered on a case-by-case basis. DIN has a standard (No. 19045-1) that includes some guidelines for typical applications in meeting rooms, which can also be applied for control rooms.

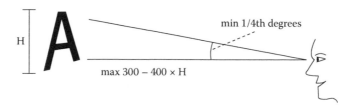

FIGURE 4.11   Image size and viewing distance.

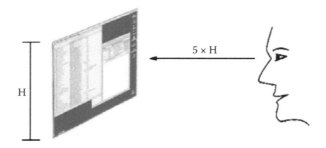

FIGURE 4.12   Ratio of image height and viewing distance.

DIN 19045-1:
Minimum viewing distance: 1.5 × width of the display
Maximum viewing distance: 6 × width of the display

A viewing distance of 4 times the height of the image is recommended as providing the optimum information from an image. The viewing distance should not be less than twice the image height, and the maximum viewing distance not be beyond 8 times the image height. It should be remembered that the DIN standard was developed before PC-based applications became state-of-the-art. Where the application includes standard PC presentations from a Windows-based PC running XGA (1024 × 768) resolutions, the longest viewing distance should not exceed 6 times the image height. If the resolution is increased to SXGA (1280 × 1024), viewing distance should not exceed 4.5 times the image height. High-resolution computer graphics requires the closest viewing distance, whereas video images can be perceived at a longer distance (depending on their size).

4.4.5   AMBIENT LIGHT

The ambient light is basically all light other than the light being directly emitted from the display (Figure 4.13).

Typically, the ambient light level in a control room will be a combination of natural light entering the room from windows, lighting in the ceiling, and the light emanating from the display itself. Depending on the reflectivity of the various surfaces in

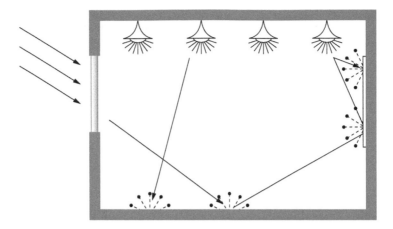

FIGURE 4.13   Sources of ambient light.

the room, the resulting ambient light level will be a combination of all these sources. The ambient light level combined with the actual luminances in the field of vision determines the average adaptation level and thus influences the viewer's perception of the display. Controlling the ambient light is essential for achieving high-quality images. Glare is the most important factor that might impede an operator's view of the display. Daylight and particularly direct sunlight are extremely powerful compared with artificial lighting, and especially compared with the power of projection systems. Direct sunlight is a negative factor for every type of display.

4.4.6   Controlling Daylight

Controlling daylight—using blinds or sun shading—must have a very high priority in any control room design. In applications where daylight cannot be controlled, consideration should be given to placing the display carefully to avoid high luminances, such as windows in the field of vision. Furthermore, direct light from the windows or the lighting onto the display should be avoided in any case, as this will affect the contrast and therefore the 'viewability'.

4.4.7   Image Quality

The perceived quality of an image is the end result of a long chain of different components and factors (Figure 4.14).

A typical chain determining image quality might include: the quality of the signal source (such as the graphic adaptor in a PC or a video camera); all of the electronic components for signal processing, storing, switching, and so forth; all the cables in a system; the display device itself; and the 'output' from the display as 'controlled light' entering our perception system via the optic nerves in the eye. The display device itself is critical to image quality, combining factors such as resolution, sharpness, colour reproduction, uniformity, contrast ratio, pixilation (in the case of

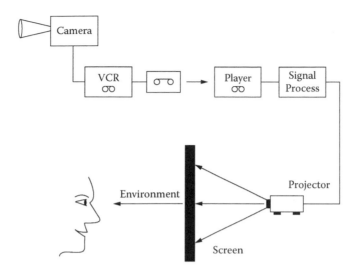

FIGURE 4.14   Chain of components affecting perceived image quality.

pixel-based display devices, such as liquid crystal display (LCD) projectors, line structure (in the case of CRT-based display devices), and dynamics (coping with moving images).

4.4.8   THE WEAKEST LINK IN THE CHAIN DECIDES THE PERCEIVED RESULT

Budgetary limitations should not influence the choice of any link in the chain. In other words, attempts to save money could result in a dramatic decrease of the quality and reliability of the different components. If this is the case then there is likely to be a reduction in image quality.

Maintaining image quality depends on several key factors. The amount of information can be adjusted according to purpose of the display. This means that the image and data content displayed on the screen can be made suitable to the needs of the operators. The designers of the control room can aim to optimise the balance between the size of the screen and viewing angles possible by the size and configuration of the room. Similarly, if possible, efforts should be made to control the field of view. Again this is a key feature of room design and layout, bearing in mind structural constraints and operational needs. It is possible to optimise image brightness according to ambient light level and also to optimise both the signal source and the display.

4.5   DISPLAY TECHNOLOGIES

Depending on the requirements and the application, there are several technologies available that can be used in control room environments. Only the more advanced technologies will be covered here. The old-fashioned technologies such as monitor walls consisting of a matrix of CRT monitors or cubes will not be explained in detail.

4.5.1 CATHODE RAY TUBE (CRT) MONITOR WALL

Disadvantages:

- High energy consumption
- Inflexible display
- Large required space
- High maintenance costs
- Complexity of display
- Old-fashioned display
- Bad ergonomics
- High thermal load
- With CRT monitors, burn-in of static images
- Large spaces in between the single display units (Figure 4.15 and Figure 4.16)

4.5.2 MIMIC PANEL DISPLAY

Mimic panel display (mosaic display) was mainly used in the energy area and in process control, but this technology is not very flexible. It is also very expensive and considered to be old-fashioned. Increasingly, more and more users are changing from this technology to the more modern display technologies (Figure 4.17).

There are two types of such displays: linear displays and graphic displays. Linear displays consist of an array of lamps presented with labels for each lamp function.

FIGURE 4.15 Example of a CRT monitor wall.

FIGURE 4.16   Example of an integrated system.

FIGURE 4.17   Example of a wall display.

Lamps are normally situated in rows and columns. The label should be sufficient for the operators to identify the nature of any problems. The disadvantage of these systems is that the lack of precise location information requires the operator to have knowledge of the facility to precisely identify the problem for the responding personnel.

The other type of display is called graphic display. These consist of lamps arranged over a graphic representation of the monitored area. The lamps are normally coloured and situated to show the location and nature of the alarm. Labels can be added for more clarity. The advantage of graphic compared to linear displays is the clear presentation of the nature and location of the alarm in relation to the area being monitored. Multiple alarms can be identified more quickly because the graphic display shows their relationship. Responding personnel can usually be directed more accurately. Typically, these systems have a larger size and higher cost than linear displays.

4.5.3 PROJECTION DISPLAYS

A visual display unit (VDU) based on projection technology is composed of a projector and a projection screen. These are either custom built by a systems integrator or based on standard display units from a display provider. The characteristics of the screen and projector jointly determine the overall performance of the VDU. The screen and projector need to be matched to meet the ergonomic demands (brightness and contrast in relation to background and ambient illumination). Adequate contrast ratios can be achieved only by using controlled lighting and projection screens that absorb a part of the ambient light. If this is not done, ergonomics is sacrificed and crucial detailed information might be lost by the operator looking at the VDU.

The basic advantage of a projection system is that the image size is scaleable. This means that the VDU can easily be optimised to the actual need when designing the installation. However, the brightness and resolution of the projector might determine a certain limitation. Projection systems are often tiled in order to 'multiply' the performance of each individual projector into a larger display.

There exist basically two different projection philosophies: front projection, where the projector is placed in front of the projection screen, and rear projection, where the projector is placed behind the screen.

4.5.3.1 Front Projection Displays

Front projection displays (Figure 4.18) are typically less expensive to set up than rear projection display systems. However, the technology has a few drawbacks that need to be considered. When front projecting, the operator is prohibited from going too close to the display as this will shield the projected image from other viewers. Front projection display needs dimmed lighting to achieve a decent image contrast. The rationale for this is that because the screen itself reflects all light, the ambient room

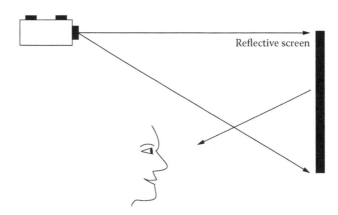

FIGURE 4.18 Example of front projection display.

light will be reflected as well. This effect 'washes out' the image and gives it a kind of 'milky' appearance.

Recently, new types of projection screens have been introduced that somehow preserve the contrast better than the traditional white screens. These new screens are either grey/dark tinted to reduce reflection or they include an integrated optical lens system. An advantage of an integrated optical lens system is that this reflects light from some directions and absorbs light from other directions. Among these 'optical front projection screens' are the dnp Supernova screen (from dnp Denmark) and the Arisawa Nexy BSB screen (Arisawa Mfg. Co., Ltd).

It should be noted that when the projector specification sheet states a certain contrast ratio, it normally refers to a completely darkened room. But if there is lighting in the room, the screen's reflection properties will normally have a higher influence on the image contrast ratio than the projector's 'theoretical contrast ratio'. A single projector typically has an image pixel resolution from 800 × 600 to 1600 × 1200 (1920 × 1080). Depending on the images being shown on the VDU and the size of the VDU, it might be adequate to use only one projector for the VDU.

However, it is often necessary to combine a number of projectors to form the required image. The projectors will typically be aligned in an array in horizontal and vertical rows. By doing this, the overall image brightness and resolution is improved. In principle, there are two methods of projector tiling: (1) soft edge projection, where each individual projected image partly overlaps neighbouring images, and brightness is reduced in the edge region; and (2) hard edge projection, where each individual projected image is aligned side-by-side with the neighbour images.

4.5.3.2 Front Projection Screen Types

The various screen types that typically are used in front projection can be divided into the following groups. (Remember: The basic physics of all the screens are that brightness and viewing angles are compromised against each other. The screen can only reflect the brightness from the projector, but the screen decides *how* it is reflected.)

- *Standard white screen*—This screen has a gain close to 1.0 and diffuses the projected light like a Lambertian diffuser (to give a similar brightness in all directions).
- *Gained white screen*—Gain is typically 1.2 to 1.8, and the viewing angles are reduced compared to the gain 1.0 screen.
- *Tinted screens*—Gain is typically 0.6 to 0.9. These screens appear grey, and they absorb a certain amount of the light and thereby maintain a better image black level.
- *Gained tinted screen*—Gain is typically 1.2 to 3.0. The screen appears 'silver-like' and while it has a good contrast ratio, it has very restricted viewing angles.
- *Optical screen*—Gain is typically 0.8 to 2.0. The contrast ratio is very high and the viewing angles can be very good, depending on the optical lens design. These screens are often limited in size, but can be tiled to the required image size.

Table 4.1 Comparison of Screen Types: Resistance to Ambient Light

Screen Type	Image Contrast Ratio	Viewing Angles	Peak Brightness
Standard white	Poor	Good	Fair
Gained white	Fair	Poor	Good
Tinted	Fair	Good	Poor
Gained tinted	Good	Poor	Fair
Optical	Very good	Good	Fair

Table 4.1 shows a comparison of screen types. Simple projectors are not appropriate for continuous operation. LCD and DLP® projectors exist, but these are designed for simple business use—for example, making short presentations.* Additionally, the lamps and components used in these projectors do not withstand comparison with the engines of cubes. The lamps and other components have a much shorter life span and, in comparison with back projection cubes, give a poorer luminance power, contrast, and image quality.

4.5.3.3   Application Sectors

This technology is used in home cinema, in office and business applications, and in conference and education settings. The technologies have a number of disadvantages including a short lamp life span of the components. Compared to the other technologies described above, these give little contrast and luminance and are sensitive to the effects of ambient light. When someone stands between the projected image source and the screen, he or she casts a shadow onto the screen.

4.5.3.4   Rear Projection Displays

In rear projection, the projector is positioned behind the screen, projecting the image onto the reverse side of the screen. The screen then transmits the projected image to the front surface of the screen and the viewers see the projected image on the front of the screen (Figure 4.19).

By its nature, a rear projection screen is a transmitting screen. It is therefore more resistant to ambient light; ambient light shining onto the screen will in principle not wash out the image but instead be transmitted through the screen. Almost all rear projection screens actually show some influence of ambient light. However, it is a fact that rear projection provides a much higher contrast ratio than front projection (typical values are 10 times higher contrast ratio than front projection). In addition to the contrast advantage, the rear projection technology also allows the operators to be located extremely close to the screen, since there is no possibility of shading of the image. One drawback of rear projection is that a certain space behind the screen is needed to house the projector. Therefore, the location of this technology needs to make allowance for the required distance from projector to screen. Often one or two

---

* DLP is a registered trademark of Texas Instruments.

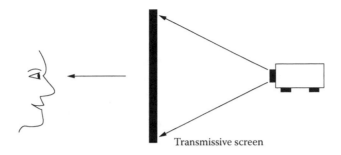

FIGURE 4.19 Example of a rear projection display.

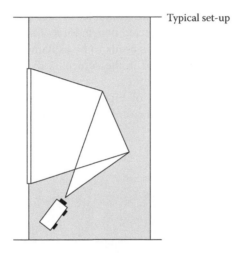

FIGURE 4.20 Projector 'folding the light path' to reduce depth.

mirrors (depending on the projector optics) are used to 'fold the light path', thereby reducing the build-in depth (Figure 4.20).

There are three basic different designs of rear projection display wall:

1. *A stand-alone device*—That is, one projector projecting onto one screen.
2. *A custom-built display wall*—A number of projectors are positioned in a mechanical structure projecting onto the screen.
3. *A cube display wall*—A number of display cubes (each containing projector and screen) is tiled to form the display wall.

4.5.3.5   Stand-Alone Devices

Since only a single projector is used, the image resolution is limited. Therefore these devices are normally not found in image sizes larger than 120 to 150 inches.

4.5.3.6 Cube Display Wall

Due to its design, the cube display wall is by far the most common projection display type. Based on standard elements, a display with any size and requirement can be established, simply by tiling the required number of cubes. The cube consists of a metal enclosure and the front surface is the screen. Mounted inside are the projector and mirror. The screen and projection engine are tailor-made to match each other with the aim of achieving crisp, high-resolution, high-contrast images with a minimum of build-in depth. The mechanics of the cube are designed to allow the cubes to be placed very close to each other to minimise the physical gap between adjacent screens. These cubes can be built up to a matrix of different displays to produce a homogenous display wall (Figure 4.21 through Figure 4.24).

4.5.4 REAR PROJECTION SCREEN TYPES

Several different types of screens are available for rear projection, and they offer very different properties in the projected image. First of all, the difference between optical screens and diffusion screens needs to be explained. An optical screen is a screen with lens elements integrated in the screen. A diffusion screen is a screen containing light dispersion particles.

The optical screen (on the left of Figure 4.25) is designed to 'catch' all the light emitted by the projector, and forward it into the viewing area in a certain way

FIGURE 4.21 Cube display wall. **(See colour insert.)**

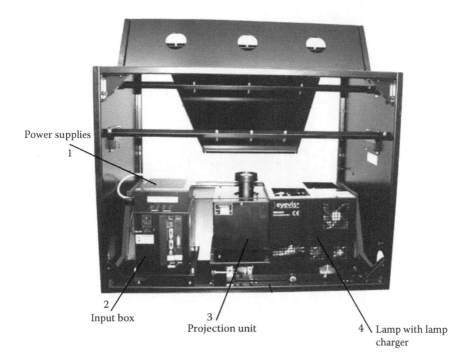

Power supplies
1

2
Input box

3
Projection unit

4 Lamp with lamp
charger

FIGURE 4.22   Inside of a cube.

(decided by the design criteria of the screen). The diffusion screen (on the right of Figure 4.25) can do nothing but diffuse the light emitted by the projector. The degree of diffusion is determined by the amount and characteristics of the diffuser particles. Diffusion screens are available in large formats (typically 300 to 400 inches) and with different optical specs (high gain and reduced viewing angles, low gain and high viewing angles, and tinted models with low gain for enhanced contrast).

Optical screens are available in limited sizes (typically 70 to 200 inches) and offer a better compromise of brightness, viewing angle, and contrast than diffusion screens. This is simply because all the light emitted by the projector 'is being utilised'. Optical rear projection screens are available in two basic different designs: single-element screens and two-element screens. The two-element screen has more surfaces that serve as lenses, and therefore it provides higher optical efficiency at the screen edges. Figure 4.26 shows that the single-element screen suffers from reflection light loss at the screen edge/corner, especially with very short projection distances.

4.5.5   Diffusion Screen Design

The performance of diffusion screens is based on the diffuser particles in the screen and the colour tint of the base material. The diffusion screens are divided into two groups:

FIGURE 4.23   Large screen cube wall installed in a curved arrangement. **(See colour insert.)**

FIGURE 4.24   Installation of a very large DLP cube wall. **(See colour insert.)**

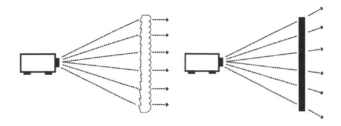

FIGURE 4.25   Optical screen and diffusion screens.

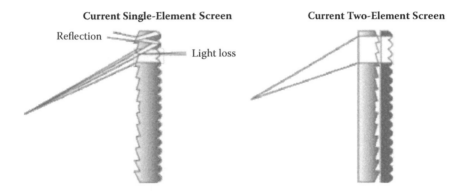

FIGURE 4.26   Single-element screen showing reflection light loss.

- *1. High gain screens (gain 2.0 or higher)* —High peak brightness and low viewing angles.
- *2. Low gain screens (gain less than 1.5)* —Low brightness and good viewing angles.

Diffusion screens can be tinted to improve the levels of black in the projected image.

4.5.6   SINGLE-ELEMENT OPTICAL SCREEN DESIGN

The screen has a Fresnel lens at the back surface, and light dispersion function on the front surface (lenticular lens or diffusion screen design). The Fresnel lens helps to collimate the light from the projector. This means that the image hot spot is reduced compared to images from a diffusion screen. This type of screen will typically demonstrate a wide horizontal and narrower vertical light distribution, or a symmetric distribution (Figure 4.27).

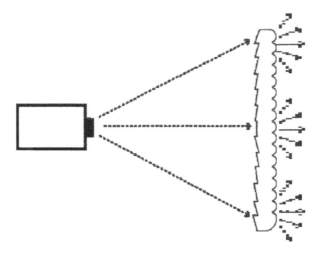

FIGURE 4.27　Projected images from a single-element optical screen.

4.5.7　DUAL-ELEMENT OPTICAL SCREEN DESIGN

In addition to the Fresnel lens element, there is a light dispersion screen element (either a lenticular, cross-lenticular, or diffusion-like screen). The Fresnel lens ensures perfect brightness uniformity across the screen surface, and the light dispersion element distributes the image into the viewing area. Designs are available both with symmetric and differentiated (horizontal and vertical) light dispersion. Two-element optical screens are available with a true contrast filter that actively absorbs the ambient light in the room, while still allowing all projected light to be transmitted (bead screens, black stripe screens). These screens show excellent resistance to room lighting (Figure 4.28).

4.5.8　SELECTION OF REAR PROJECTION SCREENS

Two-element screens are typically the most common choice for a large projection display wall. This is simply because it is important to have high contrast (readability) and good uniformity all over the display. These are also the most expensive screens to use for these applications and therefore all the screen types find their way to the display wall installation from time to time. But, of course, the overall performance of the display needs to be matched to the requirement of the application: detail readability, contrast, viewing angles/brightness, colour and brightness uniformity, and so forth (Table 4.2).

One of the most important components in a rear projection display—together with the electronics—is the screen. In order to have a perfect image, the use of a high-end screen is indispensable. Today the most widely used screen in the market of control rooms is the Black Bead screen from dnp. This screen has excellent characteristics and perfect quality for this kind of application.

FIGURE 4.1   Variations in the seeing ability of the human.

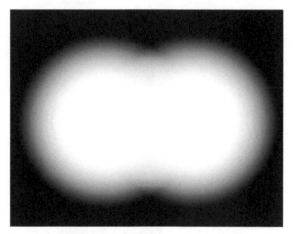

FIGURE 4.2   Perception and the human eye.

FIGURE 4.5   The colour spectrum.

(a)

(b)

FIGURE 4.6 (a) An image in 3200 Kelvin. (b) The same image in 5600 Kelvin.

FIGURE 4.21 Cube display wall.

FIGURE 4.23 Large screen cube wall installed in a curved arrangement.

FIGURE 4.24 Installation of a very large DLP cube wall.

FIGURE 4.29 DLP optical semiconductor chip.

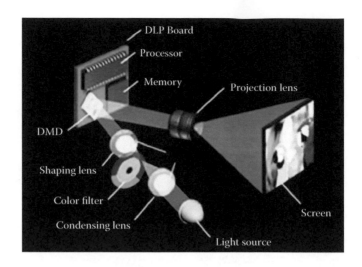

FIGURE 4.31 Single-chip DLP projection system.

FIGURE 4.32 Control room with an LCD wall, graphics controller, and wall management software.

FIGURE 4.33 CCTV surveillance control room for a city centre.

FIGURE 4.34 Simulation control room for helicopter training.

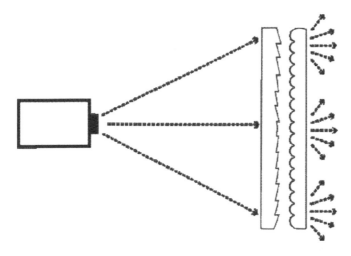

FIGURE 4.28 Projected image from a dual-element optical screen.

Table 4.2 Advantages and Disadvantages of Rear Projection Screens

Screen	Peak Brightness	Viewing Angles	Brightness Uniformity	Contrast
Diffusion (high gain)	Fair	Poor	Poor	Poor
Diffusion (low gain)	Poor	Good	Fair	Poor
Optical (single-element)	Good	Fair	Fair	Fair
Optical (dual-element)	Good	Good	Good	Good

With rear-projection cubes it is necessary to distinguish between two technologies: (1) the projection with LCD-based projection engines and (2) with DLP-based projection engines.

4.5.9 Liquid Crystal Display (LCD) Cubes

Simple projectors with *LCD technology* possess only one-colour display, which is transilluminated by the projection lamp. Because the display absorbs up to 80% of the irradiated light, the brightness of these projectors is rather poor, but they are quite cheap and are fairly light in weight. They are ideally suited for making presentations away from the office. However, if the projection takes place on a larger screen, the appropriate presentation area must be darkened.

The projection lamp is probably the most important component of the beamer. This provides the brightness, which is necessary in order to project the desired image onto the silver screen. Apart from the appropriate brightness, at the same time, the lamp must offer a pure chromatic spectrum so that the colours of the projection do not get falsified. This cannot be realised with a normal electric or halogen bulb. The lamp is a real high-tech product with the appropriate price. Most projectors contain high-pressure, gas-discharge lamps, which generate their light from a discharge arc. These are not directly attached to the current supply but get the energy from a high-frequency generator. These high-tech lights have only a very limited life span.

These projectors have applications in a number of business sectors, including entertainment, shopping malls, trade shows, and clubs. The technology has a number of disadvantages. These include: a short life span of the LCD panels, higher maintenance costs, readjustment is necessary (on a continuous basis), and the different modules lack a homogeneous image (thus giving a chessboard effect). There are also colour and brightness discrepancies within the single-projection modules. Finally, the technology gives bad colour rendering.

4.5.10 DLP Cubes

What is the DLP technology?

The DLP microchip is an optical semiconductor. Invented in 1987 by Larry Hornbeck of Texas Instruments, it is also called the digital micromirror (DMD) chip. In terms of its functionality, the DLP chip can be regarded as the most elaborate light switch in the world. This sophisticated technology is at the core of the DLP projector system. The mechanism works through an interlocking system of minute mirrors, each measuring less than one-fifth the thickness of a human hair. The DLP projection system comprises up to 2 million of these micromirrors, which are hinged to allow movement and arranged in a rectangular configuration. The projection system operates by integrating the optical semiconductor into a system comprising a digital video or graphic signal, a light source, and a projection lens. In operation, the micromirrors reflect the image onto a screen or other appropriate surface. This is called DLP technology (Figure 4.29).

4.5.10.1 The Greyscale Image

The micromirrors on a DLP chip are mounted on tiny hinges that enable them to tilt either toward the light source in a DLP projection system (on) or away from it (off), thus creating a light or dark pixel on the projection surface. The bit-streamed image code entering the semiconductor directs each mirror to switch on and off up to several thousand times per second. When a mirror is switched on more frequently than off, it reflects a light grey pixel; when a mirror is switched off more frequently it reflects a darker grey pixel (Figure 4.30).

In this way, the mirrors in a DLP projection system can reflect pixels in *up to 1024 shades of grey* to convert the video or graphic signal entering the DLP chip into a highly-detailed greyscale image.

4.5.10.2 Adding Colour

The white light generated by the lamp in a DLP projection system passes through a colour wheel as it travels to the surface of the DLP chip. The colour wheel filters the light into red, green, and blue from which a single-chip DLP projection system can create at least *16.7 million colours*. And the three-chip system found in DLP

FIGURE 4.29 DLP optical semiconductor chip. (Copyright © Texas Instruments.) **(See colour insert.)**

FIGURE 4.30 DLP optical semiconductor chip showing micromirrors. (Copyright © Texas Instruments.)

Cinema® projection systems is capable of producing no fewer than *35 trillion colours*. The 'on' and 'off' states of each micromirror are coordinated with these three basic building blocks of colour. For example, a mirror responsible for projecting a purple pixel will only reflect red and blue light to the projection surface; our eyes then blend these rapidly alternating flashes to see the intended hue in a projected image.

4.5.10.3 Applications and Configurations of a One-Chip DLP Projection System

Rear projection cubes for 24-hour operation, televisions, home theatre systems, and business projectors using DLP technology rely on a single-chip configuration as described above (Figure 4.31).

White light passes through a colour wheel filter causing red, green, and blue light to be shone in sequence on the surface of the DLP chip. The switching of the mirrors and the proportion of time they are 'on' or 'off' is coordinated according to the colour shining on them. The human visual system integrates the sequential colour and perceives a full-colour image. DLP cubes are available in various resolutions: SVGA, XGA, SXGA, SXGA+, and HDTV. Sizes range from 50 to 100 inches.

Extreme reliability, perfect image quality, and long-term stability have enabled DLP-based rear projection cubes to become state-of-the-art in control room applications.

4.5.10.4 Advantages of DLP Technology

- High mean time between failures (MTBF) of the DMD chip, up to 100,000 hours.
- Long-term stability and reliability.
- Excellent picture quality (colour, brightness, and contrast).
- Low maintenance interventions, requires few readjustments.
- No burn-in or image retention of static pictures.
- Digital technology.
- The lamp has a long life.

4.5.11 THIN FILM TRANSISTOR LCD MONITORS

The designation 'thin film transistor' (TFT) already suggests that the transistors are responsible for the screen layout. They control the action of liquid crystals, which

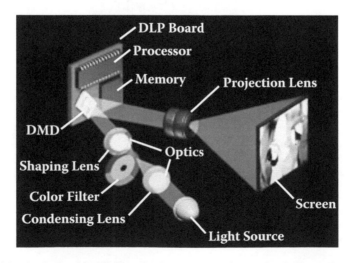

FIGURE 4.31 A single-chip DLP projection system. (Copyright © Texas Instruments.) **(See colour insert.)**

are used in an LCD. In comparison to the usual large surface flickering of conventional tube monitors, TFT monitors supply flicker-free, pen-sharp image information and therefore relieve the viewer of eye strain. Thus, even several monitors can be supervised during a longer period of time, fatigue-free, and in accordance with the strict ergonomic requirements. Additionally, TFT monitors emit almost no radiation and their images cannot be distorted by magnetic fields. Flat screens are very well suited for installation in furniture and/or rack systems because of their low heat development. However, after continuous operation, so-called ghost-images can appear very often, which are similar to the burn-in effect on plasma screens mentioned above. These ghost-images can be revoked by switching the screen off for 24 hours (Figure 4.32).

These monitors can be used in a number of applications including flight information displays on airports, trade shows, and video surveillance. They can also be used in a number of office applications. The technology has a number of advantages. For example, the product needs less space to set up and operate. The price is attractive for the level of technology. In terms of the quality of viewing, the product offers high brightness with high contrast. The disadvantages are several. In certain situations the memory effect with static images can be lost. The image format is normally 16:9 whereas the usual format in industrial applications is 4:3 or 5:4. There can be large gaps in between the single displays when these are put together. The present technology has limits in maximum diagonals.

FIGURE 4.32 Control room with an LCD wall, graphics controller, and wall management software. (© eyevis-guillaume czerw. With permission.) **(See colour insert.)**

FIGURE 4.33 CCTV surveillance control room for a city centre. (© eyevis-guillaume czerw. With permission.) **(See colour insert.)**

4.5.12 LCD Monitor Wall Narrow Bezel

The disadvantage of this technology is that the image format (Figure 4.33) is normally 16:9 whereas 4:3 or 5:4 is usually used. Present-day technology means that there are limits in maximum diagonals. There is also a memory effect with static images. The advantages are: a need for less space in which to set up and operate the equipment, and an attractive price for the quality of the technology. The technology also offers high brightness coupled with high contrast of images. This means that this technology is suitable for video surveillance tasks (Figure 4.34).

The strengths and weakness of TFT LCD screens means that this technology is a viable alternative for use in a number of situations. For example, it is suitable for use in control rooms with moving images, such as those used in video surveillance. The technology can also be used in applications where screensavers can be used, where the systems can be switched off for between 4 and 6 hours during each 24-hour operation, or where the content on the monitors can be moved periodically to other monitors. Some manufacturers have integrated functions to reduce the memory effect, such as ventilation (heat is the main reason for problems with the memory), scheduler, screen savers (running line), inverting of pictures, and panel regeneration modes. Also with a controller and related display management software, the static images can be scheduled to be moved from time to time to another display. However, it is recommended that for professional applications such as control rooms, professional displays should be used.

4.5.13 Plasma Screen Displays

The main difference between the plasma technology and other display systems is the way that a source of light is produced in each pixel (point in the projected image). In between the flat glass panels, loaded electrodes create tiny explosions of xenon gas, which lead to an ultraviolet light radiation. This radiation creates red, green,

FIGURE 4.34   Simulation control room for helicopter training. **(See colour insert.)**

and blue phosphorus light at the back of the screen. The plasma picture is extremely clear with very high sharpness of image, which extends evenly into every corner of the screen, all without any distortion or flickering. Further advantages of this technology are a narrow depth as well as insensitivity to magnetic influences. A major disadvantage is that during a continuous 24 hours of operation there is a tendency for pictures to 'burn-in'. This effect is irreversible and leads to a lasting reduction in the quality of the plasma screen display.

The technology is suitable for home cinema, shopping malls (for use with video), and trade shows. Advantages are the wide availability and a relatively low price. Disadvantages include the propensity to burn in images during prolonged periods of continuous use, an image format of 16:9 against the prevailing industrial standard of 4:3 or 5:4, and large gaps in between the displays when these are put together.

4.5.14 SEAMLESS PLASMA SCREENS

The latest technology in this field is the so-called seamless plasma display. These displays have a minimal gap between each module when built up as a video wall. However, these systems have some limitations compared to DLP rear projection cubes. A recurrent issue with plasma screens is burn-in. This effect can be avoided only when the images are moved continuously from one module to another; otherwise, burn-in will appear. Currently, there are no plasma screen displays which do not have this effect. In plasma screen displays, the image format is 16:9, but in principle the applications used are 5:4 or 4:3. This means that on a plasma screen display the projected image content will appear stretched and distorted.

Resolution of the plasma screen is only 853×480 pixels, while the applications are mainly XGA, SXGA, or SXGA+. This means the resolution will be downscaled to the low resolution of the seamless plasma screens. With plasma screen display technology the image size is only 42 inches. The mullion between each plasma screen is 5 mm and this is not seamless.

'Seamless' plasma screens use glass screens whose surface is prone to glare. This means that there are problems with the readability because of mirror effects from the ambient light. Plasma screens have a low viewing angle and also high power consumption. The attractions of the 'seamless' plasma screen technology are the relatively cheap purchasing price, their light weight, and their elegant, slim design. Disadvantages include the level of costs needed to operate this technology. The technology has a limited and low resolution to the image format and has been found to be unreliable. With continuous operation over a twenty-four-hour period, there is the burn-in effect. The technology has a mullion gap. There is also a lack of flexibility to the displayed resources which makes it difficult to manage the displays. Plasma screens don't have automatic brightness and colour correction.

4.6   THE RIGHT DISPLAY TECHNOLOGY

As explained, each of these technologies has its application, but the technology that unifies the most advantages and that is the most flexible for control rooms are rear projection cubes based on the DLP technology. Depending on the manufacturer, these technologies have more or fewer features, integrated optimising functions, and different quality levels. The main advantage of these large screen displays based on cube technology is the high video density as well as the integrated display of video and data applications. The most important aspects that have to be considered in the selection of cubes for a control centre are: form factor, costs, resolution, brightness, display properties, projector inputs, maintenance costs, and reliability. Cube-based display walls can be used in almost every area, since they can be connected to many various video inputs, can display these signals simultaneously, and can be extended easily. Only very reliable systems come into operation, which have withstood the rigours of continuous operation 24 hours a day, 7 days a week, and 365 days a year. Such systems have been especially designed for this kind of usage. Of course those systems require more space as flat screens, but taking into consideration their advantages for the 24/7 operation, space should not be an issue.

The systems are based on the DLP technology from Texas Instruments and were developed for continuous operation. The possibilities and the features of the system are nearly unlimited. The systems offer the following advantages:

- Display of many different sources.
- Complete solution from one supplier: cube, software, controller.
- Compatibility with customer systems.
- Easily extendable, a lot of options.
- Perfect overview, high image quality.
- Flexible display, layout can be changed easily and quickly.
- Cross-platform.
- Meeting of ergonomic and architectural requirements.

- Saving potentials, space, energy, weight.
- Lower service and maintenance costs.
- Functionality.
- Low heat dissipation.
- Free positioning of the windows, scalability.
- Stable geometry.
- High contrast ratio.
- High colour stability.
- Homogenous colour representation over the whole surface.
- Homogenous brightness uniformity over the whole surface.
- Long-term stability.

There are also different options available in such cubes, such as front maintenance systems and large screen walls on rails as a space-saving solution for control
rooms with limited space. As an alternative it can be stated that LCD monitors with
narrow bezels can also be used in control room areas, where the functionalities of
these displays meet the requirements.

4.6.1 Display Systems

Visualisation technology is only a part of such a large screen system composed of a
matrix of different displays. These systems in general are composed of three different elements (Figure 4.35).

An array of DLP back projection units or a matrix of any other display technology
in connection with the appropriate split controller are capable to visualise complex
systems, video signals, overview maps, and other scenarios. Additional information
can be visualised simultaneously everywhere on the wall. All input sources can be
controlled, executed, arranged, and automated with special wall management software. Control centres where important data and procedures are surveyed require
reliable, flexible, and high-quality technology.

Large screen systems based on DLP rear projection cubes must have the following features for operation in control room environments:

- Internal split controller available.
- Several inputs available.
- Local area network (LAN) connection possibility.
- Automatic colour and brightness correction to avoid the so-called chess-
  board effect and the colour differences between the individual cubes in a
  large screen wall.
- Automatic double-lamp systems.
- Minimal gaps between the projection modules.
- Possibility to build up walls in a curved arrangement with possibility to
  chooses different angles.
- Screens with nonglare surfaces and high viewing angles.
- Low noise level.

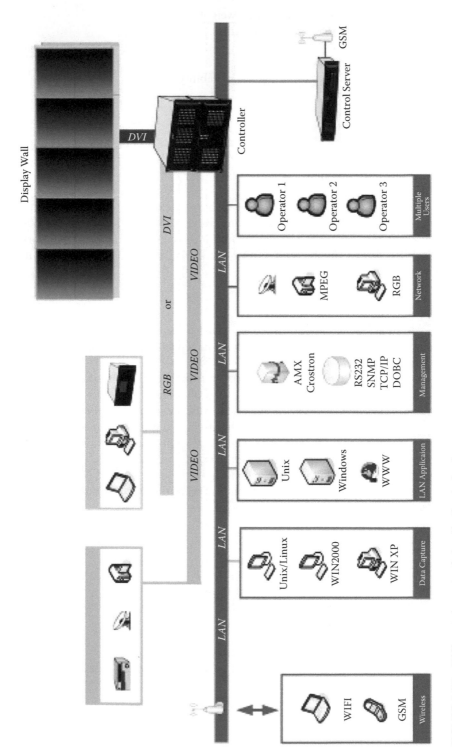

FIGURE 4.35 Different elements of visualisation technology.

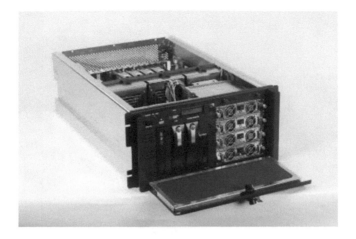

FIGURE 4.36 Inside a graphics controller.

4.6.2 GRAPHICS CONTROLLER

Graphics wall controllers (Figure 4.36) enable operators and other employees with diverse interconnected responsibilities to share a common view of images from different kinds of information sources. These sources can be remote video cameras, TV signals, workstations, and PCs, or proprietary applications. Used together with projectors, projection cubes, and other display devices, the controller then creates a huge desktop on this matrix of displays. The connected sources to the controller such as digital visual interface (DVI), RGB (red, green, and blue), or video can be inserted and placed on the video wall as required by the control room employees. So the fundamental question here with regards to choosing the appropriate controller is what has to be displayed (Figure 4.37).

Video Sources: TV signals, normally available as analogue signals (composite, S-Video, or Y/C).* This signal comes from a TV tuner, VCR or DVD player, video teleconferencing system, or closed-circuit television (CCTV) cameras. Some control rooms require only one video signal, for example, a weather forecast channel. Other applications such as traffic control or security need hundreds of video sources. Other important points here are how many sources have to be displayed simultaneously, how many in a single display, and whether the images have to be increased in scale or downscaled and by what rate.

RGB or DVI Sources: These are signals that are in general progressively scanned and do not follow TV standards. That means the complete image is refreshed in a single pass and not interlaced. Typical RGB or DVI sources are computer outputs (from VGA to UXGA). RGB or DVI sources can be captured and shown as a window on the large screen wall. These computer signals can be increased in scale and displayed on the complete large screen wall for a better overview (in general, for grid overviews for energy distribution). For these computer sources it is important that a

---

* Y refers to greyscale; C to colour.

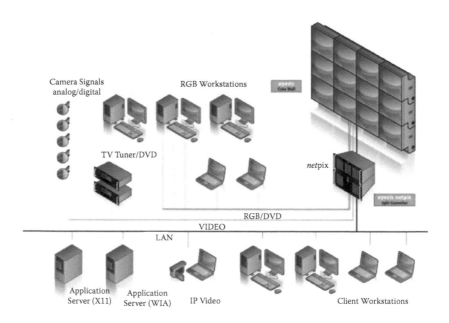

Camera Signals
analog/digital

RGB Workstations

TV Tuner/DVD

netpix

RGB/DVD

VIDEO

LAN

Application
Server (X11)
Application
Server (WIA)
IP Video
Client Workstations

FIGURE 4.37   Schematic representation of a large screen.

high-end controller with graphic cards is used with which RGB or DVI sources are displayed in real time without any time lag or delay.

Applications Directly on the Controller: Nowadays these controllers are based on the latest hardware PC technology and are very compatible with operating systems based on Windows or Linux, which makes it possible to run the customer applications directly on the controller and display it on the large screen wall. Here the main questions are whether compatibility with the operating system is given, and if there is sufficient central processing unit (CPU) and storage capacity.

Network Applications: Network applications can be connected very easily onto the controllers, because standard graphics controllers are equipped with one or several LAN inputs (standard 1 Gb). Also, the operating compatibility is the most important point.

Capture Sources from Network: Some vendors provide special management software for their controllers to enable the capture of sources in the network. It is unimportant whether these sources are from Unix, Linux, Windows, or elsewhere. Through the network they can be displayed on the large screen system. In that way it is also possible to operate one PC from a remote PC via the network.

In many control rooms, several operating systems (OS) are available such as Microsoft Windows and Unix/Linux at the same time. To provide compatibility, most systems use an X window system. For controllers using Linux OS, an X window user interface will be used and the X applications from other workstations can be easily displayed on the large screen system. But for Windows-based applications the display possibilities are limited. At present, most (approximately 98%) of graphics controllers in control rooms are Windows based. If the graphics controller is Windows based, the compatibility with other Windows systems in the network will be excellent. If the controller is equipped with an X server interface, there is also compatibility with Linux, Unix, Solaris, and other systems.

Streaming Video: This modern technology is growing very quickly. Streaming video are video sources over IP. These are generally in a low resolution (maximum 320×240); the standards are, for example, MPEG1-4, MJPEG. This means that these sources have to be scaled up to larger window sizes to be large enough to control them on the large screen system. Nowadays the controllers also have the possibility to build-in IP decoder boards.

4.6.3   WALL MANAGEMENT SOFTWARE

In large screen displays, the connected sources and processes must somehow be controlled comfortably and intuitively. Therefore there must be a software platform to ensure that. User-friendly software with a range of almost unlimited features and possibilities are now available (for example, the eyecon software from eyevis). Software such as this offers many possibilities for users and administrators. Such software has been developed through strong cooperation with users and is designed according to the actual and the growing needs in the control room area as exemplified in this chapter. Ideally, the software combines the comfortable possibilities of displaying and administering external video and RGB-signals as well as the visualisation of PC-signals via network. The integration of event-controlled actions allows the use of professional alarm-management solutions. It is possible to control and manage the whole large screen wall and the connected sources and processes. Alarm-releasing signals can be analysed by the software via the network or serial interface and can create predefined actions. The graphical user interface can also simplify the handling of complex network applications, video, and RGB sources on the wall. For the user, the system secures shorter introduction times and lower training expenses.

These advanced systems are likely to be based on a real client-server structure, allow a flexible expansion of the customers' desire and requirements, and can be based on a MySQL database. All relevant information in a wall management system is stored and released centrally in a database. This means that the control of the large screen walls is possible from all points in the network. All systems and clients have access to the database so that faulty operation and tedious reestablishment procedures can be avoided. The database is generally located on a separate control server or the graphics controller, which is used as the central station for the configuration and administrative control of the whole system. The graphics controller is used to visualise all data on the display units and is controlled by the control server, allowing the demonstration of all sources on the large screen wall (Figure 4.38).

A wall management system offers a transparent application of all resources independent whether these relate to network application, RGB, DVI, or video sources. A user-friendly graphical user interface (GUI) should be a standard so that the users can work very quickly with the system without long training hours. Ideal would be an appearance like that of Windows with a virtual desktop creating a working area, which allows an easy handling of all options via drag and drop.

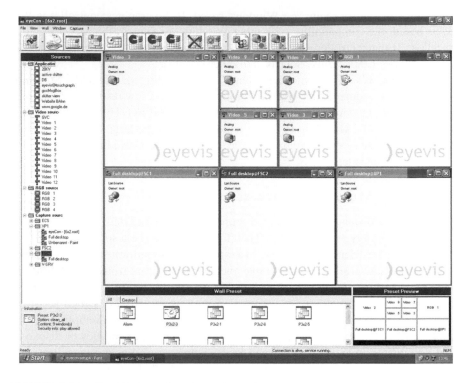

FIGURE 4.38   Graphical user interface (GUI) of a wall management system.

Below are listed examples of some basic functionalities that should be available:

- All resources can be started coevally, managed dynamically, and displayed as a window on the large screen wall.
- Administration of all services via transmission control protocol/Internet protocol (TCP/IP).
- Unlimited number of operators can use the software on their local workstations.
- Authorisations, hierarchical ranks, or ranges on the large screen wall can be freely defined for every operator.
- Definition of groups for all sources and applications.
- Manual entry of coordinates and snap-to-grid options allows an exact control of all windows.
- Zoom and scrolling options for the virtual desktop.
- Control of the wall desktop via a local keypad and mouse at the operator's workstation.
- Free positioning and scaling of all sources.
- To allow a quick switch or automation of all sources and applications, scenarios (presets) can be defined.

- Preview of all presets and easy intrusion.
- Temporal control of all presets with the option 'schedule'.
- Creation of presets with diverse options like Replace, Hide, Add, and Advanced as clear instructions.
- Intrusion of presets in certain predefined regions.
- Location over the complete large screen systems and workstations can be controlled and administrated by the wall management software.
- Via wide-area network (WAN) connection, the system enables a global management.
- Control of large screen projectors and status displays of every single projector.
- Remote control and remote diagnostics of the complete system (Figure 4.39).

In very critical applications the system and all components (graphics controller and control server) are released redundantly. Each graphics controller is connected with the cubes and each cube provides two inputs to switch the signal from the graphics controller. For example, a hot-standby system runs parallel to the normal system (master/slave configuration). In case of a failure of the master system, the slave system assumes the control and visualisation.

4.6.4   INSTALLATION OF LARGE SCREEN SYSTEMS

To ensure mechanical stability and robustness, the structures and the chassis of the cubes should have a stable and robust construction. This is the case even if large screen systems are used for fixed installations. Everything should be premounted (chassis, projectors, mirrors, screens on frame). This will save much time (and therefore money) during the screen installation. Some vendors offer systems which are delivered in preconstructed sets (the complete cube housing is dismantled in parts,

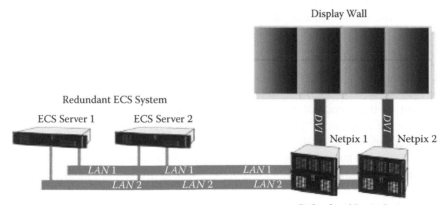

FIGURE 4.39   A redundat large-screen system.

as well as mirrors, screens, and frames). While this may seem convenient, it can also cause problems as the technology installation becomes more complicated and less accurate. Postinstallation, there will need to be adjustments to the system to ensure effective and problem-free operations.

In general, such a system consists of the following parts: a large screen cube (including 45° mirror, six-axis projector geometry adjustment, projector engine, rear panel); the screen, premounted on a screen frame; a base structure; cabling (including the connection between the controller and the cubes and/or cables to sources); a controller; source connections; and input power cabling. Before installation can begin, everything must be checked on site to see if the site is ready for such an installation. An important point to check is the accessibility to the site for the installation engineers (including room to manoeuvre the installation equipment and the components of the large screen system). The site should be checked for freeness from dust and, if deemed necessary, cleaned *before* the start of the installation process. It is wise to check availability of all source connections and power cabling (including that there is sufficient length of appropriate cabling and that the cabling, when connected, will be inconspicuous). These preinstallation tasks completed, the metal base fittings for the cubes must be fixed on the floor. This can be to the computer floor or to the base floor of the room. The standard height of such a basement is in general 1.00 metre from image bottom. Next, the installation engineers can begin mounting the display cubes.

The cubes are placed one on another and next to each other and then attached to each other. Depending on the configuration, the size of the cubes, and number of rows, it may be necessary to stabilise the base in order to prevent the completed wall from tipping over. Once the cube fixture is stable, then the screens can be inserted. The screens will need to be adjusted mechanically and fixed firmly. The controller can next be installed. Most often this will be placed inside the base of the wall (usually this is in standard 19-inch racks). Finally, the system can be connected and the sources can be plugged in. Once the mechanical installation is complete, the adjustment and the software installation can begin. After this installation of the large screen wall, the base structure can be covered and embedded in the control room design.

4.6.5 Screen Mullion

A gap between two adjacent cubes should not exceed 0.3 mm per screen, which leads to a gap between two adjacent screens of about 0.68 mm. Then the gap can hardly be seen by the users and we can speak about a seamless display. A smaller gap between the cubes is not possible because of the screen characteristics. It can also be dangerous for the quality of the large screen system. The screen is sensitive to temperature and humidity variations, so in general those systems should be operated in stable environmental conditions (air-conditioned). If, for example, the air-conditioning fails, this can cause a growing or shrinking of the screen; a gap is required to accept these variations. Also it is important that these screens are fixed with flexible springs that can accept those variations. Some vendors offer fixed solutions; this is not ideal because the variations cannot be accepted with such a solution. A gap that is too small and unable to accept the variation in the size of the screen can cause the screen to shape like a bow. Not only will this create a distorted image, but there is a risk that the screen may fall out. Such phenomenon will not occur when the screen is operated in stable climatic conditions with stable air-conditioning.

4.6.6   ENVIRONMENTAL ASPECTS

Graphic displays should be placed out of the way of direct sunlight or other strong ambient lighting. The displayed information is prone to becoming 'washed out' and hard or impossible to read from certain angles. A good location of a large screen system is just below the ceiling level and angled down to the operator's position. This location uses space that would normally be unusable. The down angle provides good visibility, does not interfere with the control room lighting, and is out of the way of direct sunlight.

A reliable large screen system requires a stable temperature and humidity control. This is important for the electronics and also the screens. Screens can be affected by changes of temperature and humidity. Ideally, they should be operated only in an air-conditioned control centre. Nowadays this tends to be standard. Another important factor is the noise created by large screen systems. Modern systems have very low noise dissipation. When a large screen system is integrated in the control centre design, care should be taken that the noise generated by the electronics is not transmitted to the control room. This can be avoided by installing sufficient levels of insulation. In modern control centres the controllers as well as other PCs tend to be located in a separate and secure server room.

As with any piece of electronic equipment, the ambient conditions are very important. Therefore it is important to ensure that such systems are installed in rooms with stable conditions. The environment must be relatively dust-free. Some vendors offer completely encapsulated systems with overpressure inside to prevent dust from entering. This is very useful for areas with a lot of dust, for example, in coal mines or in sandy areas, such as countries in the Middle East.

4.6.7   SERVICING A DLP CUBE WALL

The service of such as system is quite easy. It consists of cleaning the system so that it is free from dust, and making readjustments to the system itself. It is normally sufficient to service the equipment one to two times per year. In case of a lamp failure, the systems supplied by most vendors have automatic double-lamp systems. In the event of a lamp failure, the system switches automatically to a second (backup) lamp. It should be possible to change the broken lamp during working operations. This task is quite easy and trained operators should be able to do this by themselves.

GLOSSARY

Aspect Ratio: The ratio of the display width to the display height. The aspect ratio of a traditional television is 4:3, but there is an increasing trend towards the 16:9 ratio typically used by consumer large screen, high-definition televisions.

Brightness: The amount of light emitted from a display. It is sometimes synonymous with the term 'luminance', which is defined as the amount of light emitted in a given area and is measured in candela per square metre.

Burn-In: Phosphor, the illuminant in CRT and plasma screens, subject to a natural ageing process. If images are unchanging for a long time, this can lead to the image 'burning in'.

CCTV: Closed-circuit television, a television system often used for surveillance.

Contrast Ratio: Defined as the ratio of the luminance of the brightest colour to the luminance of the darkest colour on the display. High-contrast ratios are demanded but the method of measurement varies greatly. They can be measured with the display isolated from its environment or with the lighting of the room being accounted for. Static contrast ratio is measured on a static image at some instant in time. Dynamic contrast ratio is measured on the image over a period of time. Manufacturers can market either static or dynamic contrast ratio depending on which one is higher.

DCS: Distributed control system. In DCS, the controller itself controls the process. It has the control loops in the controller. It communicates through the high-speed Ethernet/LAN network. DCS is confined to a single unit or group of local units. Examples of uses: control system of all process plants, including refinery (petroleum), chemical plants, power plants, etc.

DLP: Digital light processing technology; with digital micromirror device (DMD), from Texas Instruments.

Graphics Controller, Display Processor, Display or Split Controller: A type of controller that splits up a video or computer signal into multiple part images so that they can be displayed on multiple screens (4, 9, 16, 25, ...) to create the full image again. In general, inside these controllers you have the possibility to add several different input cards to show these sources on the large screen; these inputs can be: analogue video, IP or streaming video, RGB, DVI, or SDI signals. For example, it is possible to have an integrated split controller in displays for a 2 ×2 or 3 × 3 video walls.

Memory Effect, Image Retention, Ghost-Images: Images that remain on the screen. One of the advantages of LCD displays compared to plasma is that LCD does not have the problem of permanent burn-in when displaying a static picture for a long time (for example, a company logo that is always present at a particular corner on the screen). But LCD displays also suffer from a similar type of image retention. The image retention that may appear on LCD displays is called memory effect or ghost-images; it is not permanent and can be reversed. Also LCD displays are not so sensitive to this effect; like on plasma, this can occur just after a few hours of use. It occurs on LCD displays when ionic contaminants in the panel migrate to the surface (usually caused by electromagnetic interference) and accumulate to areas in the panel where a static image is displayed. This will cause a drop in the drive voltage in that area and the static image remains visible even after the particular image is changed. Once the image is changed, the impurities will, with time, migrate out of the area and the memory effect should disappear. The time it takes for the image retention to disappear depends on how severe it was in the first place, and in some cases it takes so much time and effort that for all practical purposes it could be called permanent. Switching off the LCD for 4 to 6 hours during 24-hour operations could also be a solution. Also very important to prevent this image retention is the cooling of the display.

Mimic or Mosaic Display: Panels, consisting of several small LED lights, used for illuminated mimic diagram; indication and control panels; distribution of gas, oil, steam and other liquids; different industrial process with status indication; control and signalling systems of railways.

Pixel: Picture element, using the abbreviation 'pix' for 'picture'; a single point in a graphic image. Each such information element is not really a dot or a square, but an abstract sample. With care, pixels in an image can be reproduced at any size without the appearance of visible dots or squares; but in many contexts, they are reproduced as dots or squares and can be visibly distinct when not fine enough. The intensity of each pixel is variable; in colour systems, each pixel has typically three or four dimensions of variability such as red, green, and blue, or cyan, magenta, yellow, and black. Standard display resolutions include:

- VGA 0.3 Megapixels = 640 × 480
- SVGA 0.5 Megapixels = 800 × 600
- XGA 0.8 Megapixels = 1024 × 768
- SXGA 1.3 Megapixels = 1280 × 1024
- SXGA+ 1.4 Megapixels = 1400 × 1050
- UXGA 1.9 Megapixels = 1600 × 1200
- HDTV 2.1 Megapixels = 1920 × 1080
- QXGA = 2048 × 1512
- QSXGA = 2560 × 2048
- Quad HDTV 8.3 Megapixels = 3840 × 2160

Plasma Display: Plasma display panel (PDP), a type of flat-panel display nowadays commonly used for large TV displays (typically above 37" or 940 mm). Many tiny cells located between two panels of glass hold an inert mixture of noble gases (neon and xenon). The gas in the cells is electrically turned into a plasma which then excites phosphors to emit light. Plasma is, in general, used in the home environment.

Resolution: The number of pixels in each dimension on a display. In general, a higher resolution will yield a clearer, sharper image. It is normally expressed in pixels.

Response Time: The time it takes for the display to respond to a given input. For an LCD display, it is defined as the total time it takes for a pixel to transition from black to white, and then back to black. A display with slow response times displaying moving pictures may result in blurring and distortion. Displays with fast response times can make better transitions in displaying moving objects without unwanted image artefacts.

SCADA (Supervisory Control and Data Acquisition): Systems that are typically used to perform data collection and control at the supervisory level. Some SCADA systems only monitor without doing control; these systems are still referred to as SCADA systems. The supervisory control system is a system that is placed on top of a real-time control system to control a process that is external to the SCADA system. In SCADA systems, the operator/supervisor collects alarm/event data from the remote terminal unit (RTU) or programmable logic controllers (PLCs) along the service line. The operator analyses the data (alarms and SOEs [sequence of events]) and takes action if required. The RTU/PLC is the local control system that works as per the command received from the supervisory system, except for some specific control decisions (fire fighting, emergency shutdown) taken locally by the RTU/PLC. The communication path is wireless, GSM tech. (Global System for Mobile Communications), etc. Examples of uses for control are: water supply pipelines, gas/petroleum pipelines, and so forth.

TFT-LCD (Thin Film Transistor–Liquid Crystal Display): A variant of liquid crystal display (LCD) which uses thin film transistor (TFT) technology to get better image quality. TFT-LCD is one type of *active matrix* LCD, though it is usually synonymous with LCD. It is used in televisions, flat-panel displays, and projectors.

Viewing Angle: The maximum angle at which the display can be viewed with acceptable quality. The angle is measured from one direction to the opposite direction of the display, such that the maximum viewing angle is 180°. Outside this angle the viewer will see a distorted version of the image being displayed. Definitions of what is acceptable quality for the image can be different among manufacturers and display types. Many manufacturers define this as the point at which the luminance is half of the maximum luminance. Some manufacturers define it based on contrast ratio and look at the angle at which a certain contrast ratio is realised.

REFERENCES AND FURTHER READING

Berns, T., and Herring, V. (1985). *Positive versus Negative Image Polarity of Visual Display Screens*, Ergolab report 85:06. Stockholm: Ergolab.

Fenton, R.E., and Montano, W.B. (1968). An Intervehicular Spacing Display for Improved Car-Following Performance. *IEEE Transactions on Man-Machine Systems* 9, 2 (June), 29–35.

Grove, Andrew. (1990). The Future of the Computer Industry. *California Management Review* 33, 1, 148–61.

IBM. (1979). *Human Factors of Workstations with Display Terminals*, G 320-6102. New York: International Business Machines.

Schaller, Robert R. (1997). Moore's Law: Past, Present and Future. *IEEE Spectrum* 34, 6, 52–59.

Songer, A.D., and North, C. (2004). Multidimensional Visualization of Project Control Data. *Construction Innovation* 4, 173–90.

Stark, L., and Ellis, S. (1981). Cognitive Models Direct Active Looking. In Dennis F. Fisher, Richard A. Monty, and John W. Senders (eds.), *Eye Movements: Cognition and Visual Perception*. Hillsdale, NJ: Erlbaum Press.

Stevens, S.S. (1975). *Psychophysics*. New York: John Wiley & Sons.

Stone, P.T., and Groves, S. (1968). Discomfort Glare and Visual Performance. *Transactions of the Illumination Engineering Society* 1, 10–19.

Van Laar, Darren, and Deshe, Ofer. (2007). Color Coding of Control Room Displays: The Psychocartography of Visual Layering Effects. *Human Factors* 49, 3, 477–90.

5 Design of Controls

Toni Ivergård and Brian Hunt

CONTENTS

5.1 INTRODUCTION

This chapter describes the design of the more traditional controls and of specific controls for communication with computers. Traditional control panels have the advantage that they give feedback to the operator of the manoeuvres which have been carried out. It is important to supplement the keyboards with some additional types of controls that give feedback to the operator. Alternatively, the current control situation can be displayed to the operator in other ways, for example, by presenting the control state on the visual display unit (VDU). This chapter presents examples of the advantages and disadvantages of each type of control and design recommendations for these controls.

Keyboards are the traditional input devices used in data processing (DP) applications to communicate with computers. In the process control situation, it can often be advantageous to use other types of control mechanisms, such as multiway joysticks or light pens. The advantages and disadvantages of each type of control are described.

5.2   FUNCTIONAL ASPECTS OF CONTROLS

The control device is the means by which information on a decision made by a human operator is transferred to the machine. The decision may, for example, be taken on the basis of previously-read information devices, on the basis of information from other sources, or from some form of cognitive process.

Functionally, controls may be divided into the following categories:

1. Switching 'on' or 'off', 'start' or 'stop'.
2. Increase and reduction (quantitative changes).
3. Spatial control (e.g., continuous control upwards, downwards, to the left, or right).
4. Symbol/character production (e.g., alphanumeric keyboards).
5. Special tasks (e.g., producing sound or speech).
6. Multifunction (e.g., controls for communicating with computers).

Examples of control types: (1) include the starting or stopping of motors, or switching lamps on or off. Control types (2) may consist of an accelerator pedal to increase and reduce the flow of fuel to the engine. Traditionally the best-known example of spatial control (3) is the steering of a car. Examples of character production (4) include typewriting and telegraphy. Different forms of controls (5) are used for the production of sound. Of special interest here are the machines that are beginning to appear for the production and replication of human speech (see Section 3.6).

Of particular importance in control room design are the types of controls (6) used in conjunction with computers. Controls operated by hand are of particular use where great accuracy is required in the control movement. Hands are considerably better at carrying out precision movements than feet. Where a very high degree of accuracy of movement is required, it is best for only the fingers to be used.

Other alternatives are also possible for the design of special controls. For example, if one has a large crank, or two cranks coupled in parallel that have to be controlled by both hands, very fine control movements can be made.

Because the power available from the leg is considerably greater than that from the arms, foot controls are suitable for manoeuvring over long periods or continuously. Foot controls are also valuable where very large pressures are needed, as the body weight can be added to the force of the strong leg musculature. It may also be necessary to use foot control devices where the hands are occupied in other tasks. However, it should be noted that it may be necessary to use hand controls where there is insufficient space to accommodate foot controls.

When designing traditional types of controls, it is possible to design them in such a way that they naturally represent the changes one wishes to bring about in the process. For example, a lever that is pushed forwards may determine the forward direction of movement of a digger bucket. Or the flow in a pipe can be stopped by turning a knob that lies on a line drawn on the panel. In this way the design of the control increases the understanding of the current state of the process.

For communication with computers, the keyboard is often chosen for carrying out all the different control functions. Technically, it is often easy to connect a keyboard to a computer system. Other control devices also exist for communicating with computers, such as light pens. However, a particular failing of this type of multipurpose control device is that the control movements in themselves have no natural analogy with the changes that they aim to bring about in the process.

5.3   ANATOMICAL AND ANTHROPOMETRICAL ASPECTS OF CONTROL DESIGN

Some of the principal anatomical and anthropometric aspects will be considered. For detailed specifications, some excellent handbooks are to be recommended (for example, Morgan et al., 1963; Grandjean, 1988). One important limitation of the recommendations available today is that they are based on the body measurements of Caucasians. For other ethnic groups, the measurements must be adapted for their proportionally shorter or longer leg lengths.

The following rules can be applied in the design of all types of control:

1. The maximum strength, speed, precision, or body movement required to operate a control must not exceed the ability of any possible operator.
2. The number of controls must be kept to a minimum.
3. Control movements which are natural for the operator are the best and the least tiring.
4. Control movements must be as short as possible, while still maintaining the requirement for 'feel'.
5. The controls must have enough resistance to prevent their activation by mistake. For controls that are only used occasionally and for short periods, the resistance should be about half the maximum strength of the operator. Controls that are used for longer periods must have a much lower resistance.
6. The control must be designed to cope with misuse. In panic or emergency situations, very great forces are often applied, and the control must be able to withstand these.
7. The control must give feedback so that the operator knows when it has been activated, even when this has been done by mistake.
8. The control must be designed so that the hand/foot does not slide off or lose its grip.

Table 5.1 gives a summary of the areas of use and the design recommendations for different controls. The controls are discussed in more detail later. Figure 5.1a, Figure 5.1b, and Figure 5.1c give the optimal areas for the different controls.

TABLE 5.1
Recommendations for Controls

| | Stepwise Adjustments | | | | Continuous Adjustments | | | | |
	Rotary Switch	Hand Push-Button	Foot Press-Button	Toggle Switch	Small Wheel	Wheel	Crank	Pedal	Lever
Large forces can be developed	—	—	—	—	No	Yes	No	Yes	Yes
Time constraint for adjustment	Medium	Fast	Fast	Very fast	—	—	—	—	—
Recommended number of positions	3–24	2	2	2–3	—	—	—	—	—
Space requirements for placing and using	Medium	Small	Large	Small	Small to medium	Large	Medium to large	Large	Medium to large
Activation by accident	Small	Medium	Large	Medium	Medium	Large	Medium	Medium	Large
Limits of control movements[a]	270°	3.2 × 38 mm	12.7 × 100 mm	120°	None	± 60°	None	Small[a]	± 45°
Legibility	Good	Acceptable	Bad	Acceptable	Bad[d]	Acceptable	Acceptable	Bad	Good
Visual identification of control position	Acceptable	Bad[b]	Bad	Acceptable	Acceptable[c]	Bad to acceptable	Bad	Bad	Good
Checking control position on panel together	Good	Bad[b]	Bad	Good	Good	Bad	Bad	Bad	Good
Usability as part of a combination of controls	—	Good	Bad	Good	Good	Good	Bad	Bad	Good

[a] The exception is "cycle" pedals, which have no limit.
[b] The exception is when the control is back-lit and the light goes off when the control is activated.
[c] Only usable when control cannot be turned more than one revolution. Round wheels/knobs must be marked.
[d] Assuming that control can be rotated more than one turn.

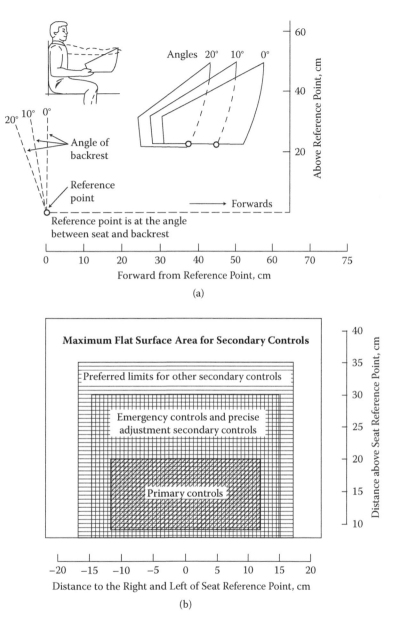

(a)

(b)

FIGURE 5.1 (a and b) Preferred vertical surface areas and limits for different classes of manual controls. (Modified from McCormick and Sanders, 1982. With permission.)

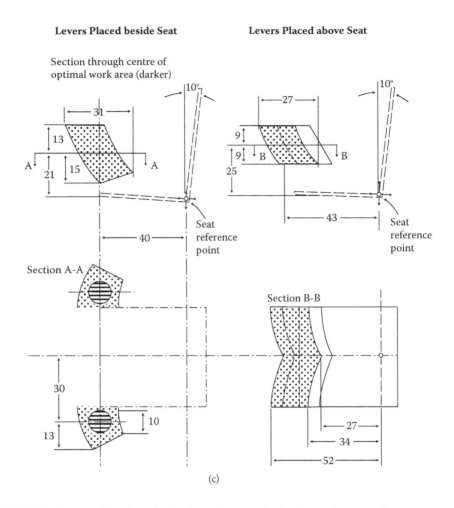

FIGURE 5.1   (continued) (c) Optimal working area for hands moving controls.

5.3.1 PRESS-BUTTONS AND KEYS

Press-buttons and keys are suitable for starting and stopping and for switching on or off. This type of control is also suitable for foot control, where it should be operated by the ball of the foot.

The following recommendations apply to both hand- and foot-operated controls:

1. The resistance of the push-button should increase gradually and then disappear suddenly to indicate that the button has been activated.
2. The top of the button should have a high coefficient of friction to stop the fingers or feet from sliding off (see Figure 5.2). Where press-buttons are to be activated by the fingers, the concave form is preferable.
3. In order to indicate that the button has been activated, a sound should be emitted if the workplace has low light levels.

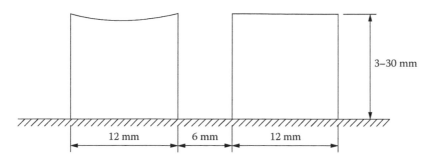

FIGURE 5.2 Press-buttons should be designed so that the fingers will not slide off them. The surface can be made concave or with some form of increased friction.

TABLE 5.2
Recommendations for the Design of Press-Buttons

	Diameter (mm)		Travel (mm)		Resistance		Distance between Push-Buttons (mm)	
	Min	Max	Min	Max	Min	Max	Min	Max
Finger	13	*	3.2	38	280 g	1130 g	13	50
One finger at random	—	—	—	—	—	—	6.5	25
Different fingers in random order	—	—	—	—	140 g	560 g	13	13
Thumb, palm	19	—	3.2	38	—	—	—	—
Foot								
Normal	13	—	13	—	1.8–4.5 kg	—	—	—
Heavy shoes	—	—	25	—	—	9 kg	—	—
Stretching ankle	—	—	—	64	—	—	—	—
Leg movement	—	—	—	100	—	—	—	—

Notes: *, not relevant; —, no information

Table 5.2 gives detailed recommendations for push-buttons.

5.3.2 Toggle Switches

Toggle switches can be used to show two or three positions. Where there are three positions, one should be up, the middle one straight out, and the other one downwards. Toggle switches take up very little room.

The following recommendations apply to toggle switches:

1. A sound should be heard to indicate activation of the switch.
2. If a number of switches are used, they should be placed in a horizontal row. Vertical positioning requires more space in order to avoid accidental operation.

Table 5.3 gives detailed design recommendations for toggle switches.

TABLE 5.3
Recommendations for the Design of Toggle Switches

Variables	Minimum	Maximum
Size (mm)		
Toggle switch		
—for fingers	3.2	25
—for hands	13	50
Travel (degrees)		
—between positions	30°	
—total travel		120°
Resistance (g)	280	1130
Number of positions	2	3
	Minimum	**Desirable**
Distance between control (mm)		
One finger—random	19	50
One finger—in order	13	25
Different fingers, random or in order	16	19

5.3.3 ROTARY SWITCHES

These can be divided into two categories: cylindrical and winged. The primary difference between these is that the winged version has a pair of 'wings' above the cylindrical part. The wings function both as a positional marker and as a finger grip. Rotary switches may have from three to twenty-four different positions. They require a relatively large amount of space because the whole hand has to have room to turn around the switch. Where multiple position switches are used, these take up less space than the number of push-buttons or toggle switches required to fulfil the same function. Rotary switches can either have a fixed scale and moving pointer or a moving scale and fixed pointer. A variant on the moving scale is to have a window that only shows a small part of the scale. Various models are shown in Figure 5.3.

The following recommendations apply to rotary switches:

1. In most applications, rotary switches should have a fixed scale and a moving pointer.
2. There should be a detent in every position.
3. The turning resistance should steadily increase and then suddenly decrease as the next position is approached.
4. Cylindrical switches (knobs) should not be used if the resistance has to be high. In these cases, wing knobs are preferable.
5. Where only a few positions (two to five) are needed, they should be separated by 30 to 40 degrees.

Switches with moving pointers are easy to read

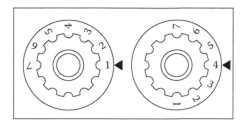

Switches with moving scales
are difficult to read

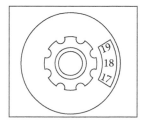

A window on a switch
with moving scale makes
it less movable

FIGURE 5.3   Design factors for rotary switches.

6. Where fewer than twenty-four positions are used, the beginning and end of the scale should be separated by a greater amount than between the different positions.
7. Where the workplace has low lighting levels, a sound should be made to denote that the switch has been activated. In these cases there should also be a definite stop position at the beginning of the scale, so that the positions can be counted out.
8. The scale should always increase clockwise.
9. The scale should not be shielded by the hand.
10. The surface of the switch should have a high coefficient of friction so that the hand does not slip.
11. The distance between panel and knob should be at least 3 mm.
12. The maximum amount of slope on the sides of the knob should be 5 degrees.

5.3.4   LEVERS

Levers are activated either by the whole hand or just by the fingers. In general, where fine control is needed, only the fingers should be used.

The following recommendations apply to levers:

1. The maximum resistance (force) for push-pull movements with one hand, with the control placed centrally in front of the body, is between 12 and 22 kg, depending on how far from the body the control is positioned.
2. The maximum resistance for push-pull movements for two hands is double that for one hand.
3. The maximum resistance for one hand moving in the left-right direction is about 9 kg, and is considerably lower in the opposite direction.
4. The maximum resistance for two-handed movements in the left-right direction is about 13 kg.
5. The lever movement should never be greater than the arm's reach without moving the body.
6. Where precision is required, a supporting surface should be provided for the part of the body used: an elbow rest for large hand movements and a hand rest for finger movements.
7. When levers are used for stepwise control (e.g., gear levers), the distance between positions should be one-third the length of the lever.
8. Where the lever also acts as a visual indicator, the distance between positions can be reduced. The critical distance is then the operator's ability to see the markings.
9. The surface of the lever handle should have a high friction coefficient so that the hand does not slip.

5.3.5   CRANKS

These are suited to continuous control where there are high demands for speed. Cranks can be used for both fine and coarse control, depending on the degree of gearing selected.

The following recommendations apply to cranks:

1. Cranks are preferable to wheels where two or more revolutions are to be made.
2. For small cranks less than 8 cm in radius, the resistance should be at least 9 N and a maximum of 22 N when rapid movement is required.
3. Large cranks of 12.5 to 20 cm radius should have a resistance between 22 and 45 N.
4. Large cranks should be used when precision is required (accuracy between a half to one revolution), with the resistance between 10 and 35 N.
5. The handle should have a high surface friction to prevent the hand from slipping.

5.3.6   WHEELS

Wheels are used for two-handed operations. Identification of the position is very important if the wheel can be rotated through several revolutions. In addition, the following recommendations apply:

1. The turning angle should not exceed ±60 degrees from the zero position.
2. The diameter of the ring forming the outside of the wheel should be between 18 and 50 mm, and should increase as the size of the wheel increases.
3. The wheel should have a high surface friction so that the hand does not slip.

Table 5.4 shows the relative advantages of different forms of control devices for computerised process systems for four common tasks.

5.4   CONTROLS FOR COMMUNICATION WITH COMPUTERS

The traditional controls in administrative computer systems are various different types of keyboards. There is often a numerical keyboard as well as the traditional typewriter keyboard. These types of keyboards have been tested over a long period of time and can be well specified. They are also thought to be well suited to most forms of administrative computer systems.

For more specialised applications—for example, computer systems for the control and monitoring of process industries—the situation is often very different. The requirement for control devices is unique to each type of process industry, depending on the process to be controlled and the type of computer system installed. It is thus impossible to give any specific guidelines for the control devices on computerised control systems. However, some overall guidelines may be given. The advantages and disadvantages of different control devices for computerised systems are examined here. Traditional control devices are rarely applied to computerised systems.

TABLE 5.4
Relative Advantages of Different Forms of Control Devices for Computerized Process Systems for Four Common Tasks

	Task			
	1	2	3	4
Control	Numeric Data	Alphabetic Data	Position Cursor	Graphic Information
Fixed-function keyboard	u	vg	nk	nr
Variable-function keyboard	vg	nk	nk	nr
Lever	nr	nr	su	nk
Wheel	nr	nr	su	nk
Light pen	nk	nk	u	nr
Electronic data board	nk	nk	nr	u
Touch screen	u	u	u	su
Mouse	nr	nr	u	su
Joystick (trackball)	nk	nk	su	nk

Notes: su, may sometimes be usable; u, usable; vg, very good; nr, not recommended; nk, the advantages and disadvantages of the application are not known.

Thus this makes it difficult to give detailed guidelines for all controls in this context. Finally, some more detailed design recommendations for the different types of keyboards will be presented.

5.5   ADVANTAGES AND DISADVANTAGES OF DIFFERENT CONTROLS FOR COMPUTERS

The more traditional types of controls can, of course, also be used for computerised systems. Controls that have been produced specifically for communicating with computers include:

1. Keyboard with predetermined functions for various keys
2. Keyboard with variable functions for the keys
3. Light pens
4. Touch screen
5. Electronic data board
6. Voice identification
7. Trackball and joystick (multiposition lever)
8. Mouse

5.5.1   KEYBOARDS WITH PREDETERMINED FUNCTIONS FOR THE KEYS

Keyboards with predetermined functions for all keys normally have two main parts: an alphanumeric or a numeric part and a function key part. The traditional keyboard—numeric or alphanumeric—will be discussed later. The function key part has different keys for different predetermined tasks, such as starting, stopping, process a, b, c, and so forth. The keyboard works by the operator pressing the keys in a certain order. The operator either remembers the order or uses some form of cribsheet. The sequence in which the keys are to be pressed is thus often predetermined both by the system and by the design of the keyboard.

This type of keyboard is characterised by the need for a large number of keys, usually one per function. Where there are many functions and several subfunctions within every main function, problems arise with grouping the keys in the proper way and in positioning the keys in a mutually logical way that is consistent in terms of movements. It is unusual to be successful with this at the first attempt; the keyboard will need to be redesigned when it has been operational for long enough to build up enough experience to determine its optimal design. Making changes to the keyboard is often costly, but if the keyboard is not redesigned at a later stage, it means that large and frequent arm movements become tiresome, time consuming, and sometimes painful. The advantage of this type of keyboard is that it needs relatively little computer programming and, to a certain extent, a standard board can be used with minimal training, at least for the alphanumeric part.

The alternative to having a large number of function keys is to have just a few of them, and to use particular codes instead that can be entered numerically or alphanumerically. This type of keyboard is best when the operator is spending a large part of his or her working time at the keyboard. However, this is relatively uncommon in process industries.

5.5.2   KEYBOARDS WITH VARIABLE FUNCTIONS FOR THE KEYS

Keyboards that have a variety of functions for each of the different keys are relatively uncommon. But these often exist as part of the more traditional keyboard (for example, the top row of keys on the keyboard). Keyboards with a variety of functions per key are often particularly useful in process industries. A common form is to have a row of unmarked keys under the monitor screen, and to have squares representing the different keys directly above them on the screen. Depending on the picture being shown on the screen, text appears in different windows showing the functions that the keys have for each frame. There are more advanced systems for this type of keyboard where there are several rows of unmarked keys, and parallel pictures are projected down from the screen onto the keyboard using an arrangement of mirrors in order to show the current function of the keys. Because considerably fewer keys are used on this type of keyboard, fewer hand and arm movements are required by the operator. This reduces the risk of errors occurring.

Another application for this type of keyboard is to build lights into the keys. The relevant keys light up for each particular function. The lights in the keys are illuminated or extinguished when particular keys are pressed, depending on the sequence of operations required. In this way the operator is guided through the correct operation sequence. Nonilluminated keys are then disconnected from the system. The risk of errors occurring with this type of system is very small, and work on this keyboard is also faster, particularly if the operator is not accustomed to the work. It is important, however, that a warning signal is produced if the lights in the keys fail.

There are also applications where keys can be pressed with different pressures. A soft (gentle) pressure on a key causes the function associated with that key to be displayed on the screen, and the action is taken if the key is pressed harder. If it is fully depressed a signal is sent to the computer dictating changes to be made to the process. The operator can also receive new information on the screen that informs him which new keys can be used.

Depending on its design, the keyboard can be preprogrammed to lead the operator naturally through the work. This type of programming of the keyboard functions may be an advantage for especially important types of operations, where errors could have serious consequences. From the perspective of the operators, a major disadvantage is that they may feel their work is being too highly controlled. Another disadvantage with this type of keyboard is that it requires a lot of programming, and this takes up a large part of computer capacity. An advantage is that the hardware does not need to be changed (rebuilding or extending the keyboard, or other changes) to any great extent even if a major change is to be made in the function of the controls. In other words, this form of control is very flexible. Keyboards can also be adapted to people with specific handicaps such as Braille scripts for vision-impaired people or special large keys for people with motor deficiency.

5.5.3 LIGHT PENS

The light pen consists of a photocell that senses the light radiated from the phosphor on a cathode ray tube (CRT) screen. The light pen reacts every time a pixel on the screen is lit up by the electron ray within the tube. The signal passes from the light pen to the computer, which at the same time receives information on where and when the spot passes different places on the screen. In this way the computer can identify where the pen is on the screen. The light pen can be used for pointing to parts of the screen one wishes to know more about. It can also be used to activate different functions. If, for example, one pointed to a valve and, at the same time, pressed a button on the side of the pen, this may cause the valve to close. The light pen is suitable for moving cursors on a screen. However, it is difficult to see any operational advantages of light pens over other controls.

If a light pen is used over a long period, it is necessary to have a specially designed armrest to prevent discomfort. The light pen has to come close up against the screen, which means that it is impossible to have any form of reflection shield or filter on the screen, and this can give rise to visual problems. Positioning of the light pen must be exact, which makes considerable demands on vision and also contributes to bad working posture in many instances.

5.5.4 TOUCH SCREENS

Touch screens involve moving the finger, a pen, a pointer, or some other object within an active matrix placed over the screen. This active matrix may be designed in several ways. It could be composed of a thin metal net which, when touched, completes an electrical circuit. Electrical bridges and infrared beams can also be used to determine touch on the screen. Another type of touch screen is based on the use of a transparent material that senses the pressure of the touch on the screen. Special measurement bridges are used to determine how the pressure field is distributed over the screen, and the position of the touch is deduced from this.

Functionally, the touch screen is very similar to the light pen and has similar advantages and disadvantages, although an additional disadvantage is that the screen becomes dirty. An advantage is that it is sometimes faster to point with the finger than with a light pen. However, the technical reliability of the touch screen is usually considered to be lower than that of the light pen.

Manually inputting information directly on VDUs has rarely been very widely applied, probably because it is often impractical and, at the same time, it might destroy the surface of the VDUs. More common are the use of light pens and other means of inputting information on the VDU screen. One advantage of a touch-screen type of system is the potential to develop tacit skills and mental models. However, there is limited information in this area. If it is necessary to transmit a large number of different types of words and information to the computer, the traditional keyboard is preferable.

5.5.5   Electronic Data Boards

Electronic data boards consist of a rectangular plate that represents the surface of the screen. Some form of electric field is created over the plate. When a sensor is run over the board's surface, it 'senses' the position of the sensor on the board. One common form of board is placed directly onto the screen, and in this case functions very like a light pen or touch screen. Another form of board is placed beside the screen, and one can work with a transparency of a picture. One of the advantages of the electronic data board is that one can very quickly make drawings or change them.

5.5.6   Voice Identification Instruments

Instruments for voice recognition and identification have been connected to computers for some time. Recognition of speech, however, is much more difficult (see Section 3.6). There are many apparent advantages of this type of device. It is, for example, very flexible and demands no special motor skill from the human operator. The problems are that the equipment available today requires specially trained operators who have to use a limited vocabulary and have to speak at a particular speed. In the future, however, this type of control may well be more widely applicable. Its present-day applications are primarily for different forms of emergency and alarm situations. There are many interesting possibilities for this form of control device within the process industries. Development should progress in such a way that natural words and sentences will be able to be used directly. In an emergency situation that requires immediate response, the operator should be able to shout 'Stop' to control the process if he or she considers this the appropriate action for the situation.

5.5.7   The Trackball

The trackball is a mounted sphere that can be rotated in all directions and can be placed on a table or special fixing. The ball is usually used for moving the cursor on a screen. The cursor moves a certain distance (x/y directions) or with a speed proportional to the movement of the ball.

5.5.8   The Joystick

The joystick, which is a lever movable in all directions, has a similar function to the trackball.

5.5.9   The Mouse

The mouse is a small device with wheels or a ball mounted on the underside. If the mouse is moved to the left or right, this represents a corresponding movement on the screen. The mouse is especially suited for moving the cursor and for transferring graphic information. There is no conclusive evidence to produce recommendations for the use of the trackball, mouse, or joystick. In practice most people seem to prefer the mouse if one has access to a free table surface; otherwise, the trackball is generally preferred.

5.5.10 OTHER TRADITIONAL COMPUTER CONTROLS

There are also many traditional types of control, such as small wheels or levers. These more traditional types of control devices are usually used for moving the cursor on the screen. However, in the future there will be a need for new types of controls which suit the computer applications within the process industries better. Traditional control panels, dynamically presented on a VDU, can sometimes be an improvement on a keyboard to provide direct feedback of different control settings.

A very good illustration of feedback related to controls is the manual gear stick in a motor vehicle. A trained user can tell its functional position from a quick glance at its position or even by touching and holding the gear knob, and by the sound of the engine. This form of feedback is partly tacit and it helps the operator to update his ongoing tacit knowledge of the current state of the process. While this situation is evident in driving a vehicle, it is not apparent when operating a computer keyboard. There is a need for new creative solutions to this problem. To be solely dependent on keyboards is insufficient as to a large extent a keyboard does not provide process feedback to the operator.

5.6   THE KEYBOARD

The keyboard is still the most common computer input device. The design of the keyboard has a significant effect on the operator's performance in terms of speed and accuracy. The most common keyboard layout is the QWERTY layout. Where two hands are used on the keyboard, 57% of the workload is on the left hand, even though 80% of the population is right-handed. This is advantageous for the type of job where the right hand alternates between handwriting and typing. It is also important to have one standard keyboard layout. Although the QWERTY layout is not the most efficient (it is said to have been designed for slowness, so that the keys on early mechanical typewriters did not become overloaded and snag), it is now the best compromise as it has become the standard keyboard.

Keyboards are usually designed so that the alphanumeric section is in the centre, with the cursor, editing keys, and numeric keypad to the right. Function keys may be placed anywhere on the keyboard, but in order to give an aesthetically pleasing design they are often placed on the left-hand side. Lateral hand movements also require less energy than longitudinal (front to back) movements.

There are no specific recommendations for keyboard layout, as their design is extremely sensitive to the task being carried out. However, a degree of flexibility must be incorporated in their design to cater for all variations in user requirements. One solution to this would be to develop a modular keyboard consisting of several units. Each unit would be made up of a different set of keys. The units could be arranged in the desired layout based on the results of the task analysis. However, care must be taken in using a flexible keyboard configuration due to the risk of a negative transfer of training. For example, if an operator carries out a number of different tasks, and different keyboards are used for each of the different tasks, then high error rates must be expected.

Chord keyboards are a combination of a keyboard and a coding system. In a similar way to keys on a piano, one can press several keys at the same time. The advantage of this type of keyboard is that key-pressing speed compared with a standard typewriter is considerably better than 50% faster. There are, however, no special design recommendations for this type of keyboard. In general, it may be said that this type of keyboard needs further research before any firmer recommendations can be given regarding suitable areas of use and suitable design. Its main application is a kind of stenographic keying which allows direct imputing of text at the same speed as it is spoken. However, this demands a rather long period of training and operators need a particular aptitude of sensor-motor skills of their hands. The application in industrialised control rooms is probably very limited; however, there might be some use for this type of keyboard in special emergency situations. In other more administrative control situations, there might be other possible areas of use. One example could be stock market trading where success depends on speed of decision and action in sending complex messages.

The numeric keyboard appears in two different designs. The accepted layout is a 3 × 3 + 1 key set, but there are two alternatives within this. Adding machines have the 7, 8, and 9 keys on the top row while push-button telephones have the 1, 2, and 3 keys at the top (see Figure 5.4). In the future, all telephones will use the 1-2-3 keypad. Once this happens, it will be recommended that all numeric keyboards are of this design. Uniformity is important, and the user should not have to switch from one keyboard design to the other (with concomitant need to shift learned skills of keyboard usage).

The height of the keyboard is largely determined by its physical design, for example, electrical contacts and activating mechanism. Thicker keyboards (greater than 30 mm thick) should be lowered into the table surface to ensure a correct user posture. Unfortunately, this does not allow for flexible workplace design. Ideally, keyboards should be as thin as possible (less than 30 mm thick from the desk surface to the top of the second key row (ASDF...) and not need to be lowered into the

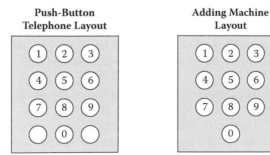

FIGURE 5.4   Layout of a numerical keyboard.

surface. Product development, particularly by Ergolab in Sweden, has resulted in thinner keyboards, and this allows for a more flexible workplace design.

Keyboards can be stepped, sloped, or dished. There is no evidence on the relative advantages of any of these profiles. The most important factor is for the keyboard to be able to be angled between 0 and 15 degrees up at the back and, if the keyboard operator is standing, it is advantageous if it can be raised at the front from 0 to 30 degrees.

The size of the key tops is a compromise between producing enough space for the finger on the key, while at the same time keeping the total size of the keyboard as small as possible. Key tops should be square and 12 to 15 mm in size. This size is quite sufficient for touch typing, but in cases where keyboards are used for other tasks—for example, on the shop floor—key sizes can be larger. The spacing of keys is standardised to 19.05 mm between key top centres. This is within the ergonomic recommendations of between 18 and 20 mm. The force required for key displacement should be the same on all keys. For skilled users, the actuating force should be 0.25 to 0.5 N, and the key displacement (travel) 0.8 to 1.0 mm (from rest to activation of system). For unskilled users the force should be 1 to 2 N and the displacement 2 to 5 mm. The user requires feedback to indicate that the system has accepted the keystroke. This is an important keyboard characteristic, although the exact requirements vary according to the individual levels of user skill.

In normal typing and other key-pressing tasks, there is kinaesthetic (muscle) and tactile (touch) feedback from the actual depression of the key, auditory feedback from the key press and/or activation of the print mechanism, and visual feedback from the keyboard or from the output display. For skilled operators, feedback from the keyboard (sound and pressure change) is of little importance. When learning, and for unskilled operators, this feedback is important. The operator should be able to remove the acoustic feedback.

The colour of the keys is not generally regarded as important. A dark keyboard with light lettering is preferable when used in conjunction with light-on-dark image displays, and care should be taken not to cause any distracting reflections on the screen by light key colours. Matt finishes should be used where possible. The recommended reflectance factors for keyboards used in conjunction with negative-image (light on a dark background) VDUs are:

1. The lettering on the keys should be light and clearly defined. Its minimum height should be 2.5 mm in good lighting conditions. In the case of function keys, certain abbreviations may be required. These must follow a clear, logical pattern and be easily identifiable by the operator.
2. Keyboards used with positive image (black on white) VDUs should be lighter in colour with darker text. All key top surfaces should have a matt finish.

Care should be taken when using colours to code various function keys. Attention should not be drawn to a red key or a group of red keys if their importance in the system is minimal. These principles concerning colour may also be applied to any information lights found on the keyboard. There are international colour standards (IEC 1975, Publication 73: *Colours of Indicator Lights and Push-Buttons*) that can apply to both keys and information lights. These standards should be adhered to wherever possible.

REFERENCES AND FURTHER READING

Gantzer, D., and Rockwell, T.H. (1968). The Effects on Discrete Headway and Relative Velocity Information on Car-Following Performance. *Ergonomics* 11, 1, 1–12.

Grandjean, E. (1988). *Fitting the Task to the Man* (4th ed.). London: Taylor and Francis.

Hunt, D.P., and Craig, D.R. (1954). *The Relative Discriminability of Thirty-one Differently Shaped Knobs*, WADC-TR54-108. Dayton, OH: Aero Medical Laboratories, Wright Air Development Center, Wright-Patterson, AFB.

IEC. (1975). *International Standard, Colours of Indicator Lights and Push Buttons*, Publication no. 73. Paris: International Electro-technical Commission.

McCormick, E.J., and Sanders, M.S. (1982). *Human Factors in Engineering and Design* (5th ed.). New York: McGraw-Hill.

Morgan, C, Cook, J., Chapanis, A., and Lund, M. (1963). *Human Engineering Guide to Equipment Design*. New York: McGraw-Hill Books.

Part III

Design of Control Rooms and Their Environment

6 Control Room Layout and Design

Toni Ivergård and Brian Hunt

CONTENTS

6.1 INTRODUCTION

This chapter begins with a description of the principles governing the location and positioning of equipment in a control room. This is followed by a discussion on the relationship between the positioning the equipment and the individual information and control devices. Amongst other things, when positioning equipment, it is important to take account of the frequency and sequence of use, the degree of importance, and the basic functions provided by a particular device. The knowledge and experience of the operator are also engaged, as the operator's expected behaviour in different situations must be taken into account.

There follows a description of the design of control and information panels, particularly with regard to the human anatomy and physiology in different working postures. The various types of supporting surfaces required in a control room, depending on the types of tasks carried out there, are also discussed. Some examples of control room design are presented toward the end of this chapter.

6.2   PRINCIPLES FOR POSITIONING OF EQUIPMENT
AND FURNITURE IN THE CONTROL ROOM

In the design and positioning of equipment and furniture in the control room it is important to use a systematic approach. It is essential for this work to be carried out in accordance with the guidelines set out in Chapter 13 and with the cooperation of all the parties affected, including architects, instrument designers and, of course, with particular reference to the operators. It is important for the operators to have the opportunity to take part in the design of their workplace. The operators, as users, are often the best experts on the physical design and positioning of equipment due to their practical experience. The experience of the operators with the equipment allows the customisation of the control room layout. In addition, new environmental or health and safety laws are giving operators new rights of participation in this work in order to protect their interests and create a good working environment.

The planning work is done in four different stages:

1. Analysis based on job descriptions related to the function of the process.
2. Determination of staffing levels, and description and specification of equipment.
3. Producing sketches and drawings (including three-dimensional drawings).
4. Making models and mock-ups.

The aim of the analysis stage is to form a basis for the designers to work on; this will be discussed in more detail later. Determination of the staffing levels and requirements is a task, which is included in the analysis. Here it is important that any characteristics of the operators, such as anthropometric measurements, should be determined and specified. The various types of equipment to be used in the control room must also be specified. The specification of this must include, for example, physical dimensions, noise level, and heat production.

Production of sketches and drawings is an important stage in the work. Ordinary two-dimensional drawings and sketches are often difficult to read, and give a very limited understanding of the final working environment. For example, it is very difficult to appreciate physical factors such as lighting or noise or to visualise a three-dimensional room from two-dimensional drawings. It is therefore important to produce additional three-dimensional drawings, which should subsequently be complemented with small or full-scale models and/or mock-ups of the proposed workplace. It is only when these have been produced that the relative advantages and disadvantages of any particular layout can be discussed with the operators.

6.2.1   ANALYSIS

A common error in designing a workplace is to produce detailed specifications immediately. It should be reemphasised that even in this connection the design process must begin with the production of an accurate analysis of the assumptions and conditions for the design work.

The analysis stage should include the following points:

1. Decide upon all the tasks to be carried out in the control room.
2. Group the tasks according to the operators, and plan the relationship between workplaces.
3. Plan information and control devices and the work surfaces in the different workplaces:
   a. Information requirements
   b. Control requirements
   c. Work surface requirements (other than those for information and control devices, e.g., writing surfaces)
   d. Resting surface requirements
4. Plan secondary functions and requirements for physical communication in the room and between rooms.

The determination of tasks within the jobs is done on the basis of the preliminary system analysis (see Chapter 13). Based on this general description of the tasks, it is possible to assign the tasks to different workers, and this is also commonly carried out at the system analysis stage. This grouping into different jobs can then be used in the creation of natural relationships between the different workplaces. The various job holders who need to contact each other should easily be able to do so. From this starting point it is then important to study the individual workplaces in more detail. This task takes into account the individual operator's anatomical, physiological, and psychological characteristics.

The workplace should be designed so that the operator has:

1. Natural and comfortable visual conditions.
2. A comfortable reach for all tasks that need to be performed. This determines the conditions for positioning of controls and other working surfaces.
3. The possibility of rest. The working posture and need for the body to rest during work are noted, and the need for resting surfaces (e.g., in the form of chairs) should be specified here.

6.2.2 Relationships between Different Workplaces

Even where the operator usually works alone, it occasionally happens that there are several other operators working in the same control room. And even an operator working alone will have several workplaces at different locations within this control room. It is important that the different workplaces are positioned in the correct relationship with one another. For the work to be done efficiently, the operators must be able to have contact with one another easily without needing to be disturbed. Teamwork should also be encouraged by the appropriate positioning of workplaces.

In order to fulfil these various aims in the siting of different workplaces in a control room, frequency and sequence analyses may be used. The best positioning of machines, instruments, and operators in relationship to each other is achieved if the various forms of connections (such as movements and messages), which are important for the system under design, are optimised. The term *connection* here means all forms of contact among operators and between operators and machines. Examples of this are that the operators need to be able to talk to each another, and they must be able to read particular instruments in order to carry out particular control movements. Connections also include moving between two places or moving material from one place to another.

Optimisation of the connections means that necessary contact is easy to 'service', which normally implies that a person with whom one has frequent contact sits close by, and someone contacted only occasionally sits farther away. Also, instruments that are used regularly should be positioned close to the operator, while less frequently-used instruments should be positioned farther away. In addition, the contacts that are important—for example, those that reduce the risk of serious faults—should be well positioned in relation to the appropriate operators and instruments.

Whether the contact for a certain connection is good or not depends to a large extent on the type of contact. If the task is one of supervision (for example, to see what another operator is doing, because their information is required for one's own work), one can sit behind, at an angle behind, or slightly above the other worker. If it is more a question of talking to each other and having a mutual interchange of information, it is best to sit opposite, or almost opposite, one another.

A link analysis is done in the following way:

1. Draw a circle for every person included in the system, and write down their function (e.g., supervisor or operator).
2. Draw a square for each part of the machine used by the operator and indicate its functions using letters or symbols.

   It is of no great importance how the circles and squares are positioned at this stage, as long as they are distinguishable (see Figure 6.1a).

3. Draw lines between the operators who have connections with each other in the system.
4. Draw lines between each operator and the machines he or she will have contact with.
5. Redraw the diagram so that the lines cross each other as little as possible (see Figure 6.1b).

For many analyses, this procedure is sufficient. It shows how the operators and machines interact, and how they can be separated where there is no contact between them and brought together where closer contact is required. Where many of the paths cross, the frequency and importance of use must be examined. This is achieved by comparing every connection using at least one of the following methods:

1. Where importance is the criterion, an experienced person who knows the different parts of the system can be used to rank all the links in order of their importance.
2. Where frequency is the criterion, information from a simulation of the system can be used in order to determine the relative frequency of the contacts, and then to rank the transactions accordingly.

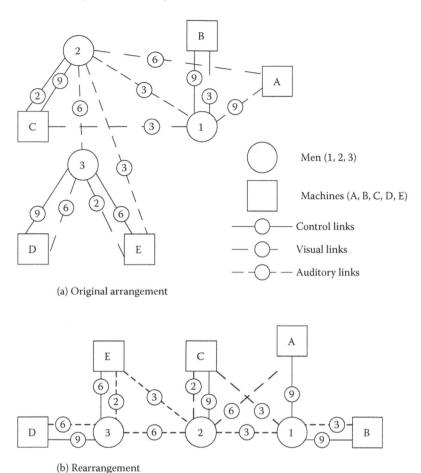

(a) Original arrangement

Men (1, 2, 3)

Machines (A, B, C, D, E)

Control links

Visual links

Auditory links

(b) Rearrangement

FIGURE 6.1 Example of the positioning of operators and machines based on the contact between them (the frequency of contacts is indicated by the figures in the smaller circles).

3. In cases where both frequency and importance are the criteria, the experienced operator can again be used to assess the various links and to rank-order them.

An excellent way of making a diagram of this type clearer is to draw thicker lines to indicate the most frequently-used or important links.

1. Redraw the diagram so that the links that are the most important/most frequently used are the shortest, and reduce the numbers of crossings. This is now the optimal link diagram.
2. Redraw the diagram again so that it fits into the proposed space, or even better, redesign the space so that it fits the diagram.
3. Determine the final diagram on a scaled drawing using the actual positioning of the operators, machines, and equipment in the system. A drawing of

this type helps in the visualisation of the physical space required. It may also be a help in detecting difficulties that were not discovered in the analysis.

6.3   PLANNING OF INFORMATION, CONTROL, AND OTHER WORK SURFACES

Planning an individual workplace starts from the assumption of natural and comfortable visual conditions. Given this, the positioning and design of information and control devices, together with the work surfaces, can be discussed in order to achieve a logically designed workplace in which it is easy to work, is not fatiguing, and allows for fast and accurate work.

When designing the instrument panel itself, the tasks to be performed by the panel must be determined first. The functions that need to be displayed using the panel must then be determined and described. A detailed description must also be made of how the information on these functions is to be received. It is important to know, for example, whether just a single instrument panel is needed, whether it is to be one of a repeated series of similar instrument panels, or equally, whether it will be just a segment of a panel that can be successively expanded. All these facts will then form the basis for the design of the instrument panel.

Various methods are used for designing the instrument panel and which one is chosen depends on what will be the panel's function. For certain applications, an instrument panel that represents a model of the process may be best. In other cases, this form of panel may be wholly unsuitable and give very unsatisfactory results. It may be better to have an instrument panel on which the instruments are positioned according to one of the following models:

1. *Frequency of use*—If instrument A is read more frequently than instruments B and C, A is positioned nearest to the line of sight.
2. *Sequence of use*—If instrument A is always used before B and C, then B and C will be placed to the right of A in the line of sight.
3. *Degree of importance*—In cases where the instrument is very important, it can be placed centrally even if not used frequently.
4. *Similarity of function*—Instruments that show the same function (e.g., temperature) or cover a particular part of the process can be positioned together into a group.

The following three points apply to all the above models. All three functions are explained below, together with design recommendations.

1. *Visibility*—The operator must be able to see all instruments from his normal workplace without abnormal movements of his head or body (see Figure 6.2). The following design principles for positioning with regard to the visibility of different instruments on a panel should be followed:
   a. *Warning signals and primary instruments*: Reading of this category of instrument must be possible without the operator having to change the position of the head or the eyes from the normal line or position.

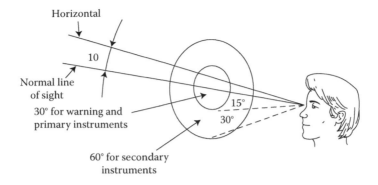

FIGURE 6.2 Positioning of visual instruments.

  b. *Secondary instruments*: Reading can be done by changing the direction of the eyes but not the position of the head.

  c. *Other (infrequently-used) instruments*: These instruments do not need to lie within the normal line of vision.

2. *Identification*—The operator must be able to find an instrument or a group of instruments rapidly without making mistakes. The individual instruments or groups of instruments should be separated in order to facilitate identification. Other ways to differentiate instruments are:

  a. Different colours* or shading between the main and subsidiary panels.

  b. Lines of different colours around the instruments.

  c. A subsidiary panel away from the main panel.

  d. A subsidiary panel sunk into the main panel.

Apart from the above, textual signs can be used. Rules governing the use of such signs are:

  a. Positioning must be consistent, e.g., either over or under the instrument.

  b. All panels, groups of instruments, and individual instruments must have signs. The size of the signs (text) should increase by 25% from smallest to largest; for example, single instruments are smallest, then groups, and so on.

  c. The text must be horizontal.

  d. The signs that are not related to function, such as manufacturer's name plates, must be positioned so that they cannot be confused with functional signs.

  e. Signs should not be put on curved surfaces.

---

* The colour red is normally used on instruments showing danger. In Chinese culture red is a 'lucky' colour and is thus not meaningful to represent danger. Obviously, in a globalised world, the colour red is used on stoplights and other 'danger' signals but in this context its use is a result of Western influences.

3. *Stereotypical behaviour*—Both control devices and instruments should be positioned and give the readings that people would expect. Human beings often have a predetermined expectation, which is based on experience, of what will happen in different situations. An example is that of a reverse gear, which should be in such a position that the gear lever has to be moved backward to engage it.

Particular changes on an instrument produce certain expectations, e.g., the clockwise movement of the pointer on a circular dial indicates an increase. Certain relationships between the movement of a pointer on an instrument and the movement of the controlling device are also expected (see Figure 6.3).

a. Learning times will be reduced.
b. If account has not been taken of the stereotypical expectations, there is a great risk that accidents will occur, together with poorer performance, in emergency situations.

During the computerisation of a process control system, there is often a tendency to remove the overall display panels (which give an overview and general description of the elements in the process, and which may also have indicators for various functions in the process). This is usually a mistake, as the display panels often fulfil

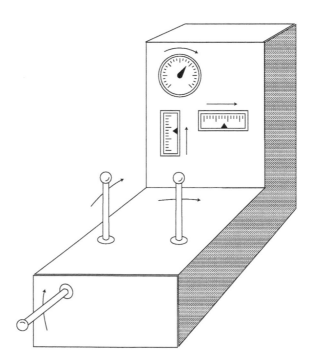

FIGURE 6.3 Expected relationships between instruments and levers; arrows represent the direction of increase.

a more important function in the computerised system than in the traditional one. The traditional panel usually gives an overview of the status and build-up of the process. Visual display units (VDUs) give a considerably poorer overview, so they need to be supplemented with general overview information. Only a limited amount of information can be shown at any one time on a VDU, so the information is generally presented in sequence. Even if the operator has access to several screens (parallel presentation), he or she is often forced to leaf through different frames on the VDU in order to build up a picture of the status of the system. This may be troublesome in situations that demand rapid action, and in some cases may be dangerous. A functional overview display panel, which has various indicators on it, and maybe also instruments, can be an important supplement here.

Figure 6.3 shows recommendations on the natural relationships between functions in the process and the information devices, and between the information and control devices. It is difficult when using computerised systems to bring about these natural relationships. The information on the VDU screen does not usually bear any logical resemblance to what is happening in the process. The control movements on the keyboard are often even less related to what is happening in the process (for example, increase or reduction should be represented by movement of a lever forwards and backwards respectively). Even in computerised systems, one should strive to maintain these relationships.

The operator should have the VDU screen within close view; normally, a keyboard or some other computer-controlled device is at his or her disposal. These controls should be placed so that the operator can position them with his or her arm in a comfortable, relaxed position. Further recommendations on the positioning of the table, work surfaces, and control panels are given in relevant sections in this chapter.

Chapter 4 presents the development of display technologies that offer many opportunities for new ways of designing dynamic overview displays. The general principles presented below give some design guidelines which are also highly relevant for the new generation of display technology.

6.4   DESIGN OF INFORMATION AND CONTROL PANELS

The information part of the control panel may have different groups of information devices:

1. Overview of information devices:
    a. Dynamic
    b. Static
2. Specific information devices:
    a. Dynamic
    b. Static

An overview information device aims to give the operator a general view of the design or status of the process. General information may either be static (for example, a flow chart painted on a wall diagram) or it may be dynamic so that it includes instruments and shows different aspects of the process. A dynamic information device may also contain symbols that change (lights that are on or off depending, for example, on whether a particular valve is open or closed).

The specific information device (usually a cathode ray tube [CRT] screen) is the one from which more detailed information is obtained, such as readings that enable control actions to be made and set values to be implemented. The specific information is often dynamic, but it may also be static, for example, in the form of information in tables and charts.

There should preferably be only two viewing distances: ordinary reading distance (ca. 30 to 35 cm) and long distance (several metres to view the overview panel, for example). When CRT screens are used, a compromise usually has to be made as the screen cannot be placed too close. There may in this case be three normal viewing distances (for people with spectacles, this means the use of multifocal lenses).

It is important that the viewing distance of the CRT screen is not less than 33 cm; preferably, it should not be less than 50 cm. Too small a distance causes great strain on the eyes and normally requires the use of spectacles even for those with normal vision. The maximal viewing distance depends on the size of the information displayed.

All the important visual information, and all the visual information that is used frequently, should lie within the normal viewing area. The maximum viewing area (angle) is that area within which the operator can see the instruments or controls relatively quickly and reliably by moving only the eyes and without needing to carry out any tiring neck or eye muscle movements.

Figure 6.4 shows an example of the recommended dimensions of the control desk for standing work. From the figure it may be seen that a normal line of sight either directly forwards or a little downwards (maximum 10 degrees) is optimal.

Figure 6.5a shows how a VDU screen should be positioned for ease and comfort of use where the majority of the work is carried out at the VDU screen, a situation that should normally be avoided. Where the VDU work is more sporadic, a flexible positioning of the screen as shown in Figure 6.5b can be recommended.

Most control room applications where VDUs are used require the use of more than one screen. It is usual for at least three to be necessary for the central workplace. These screens are used for different purposes:

1. One for overview information.
2. One for detailed information.
3. One for alarms.

It is also a good idea to have one or more screens at a different, separate workplace for planning work and as reserves.

VDU screens only give information in sequence and only a limited amount of information can be presented at any one time. The overview information as presented on a VDU cannot contain particularly large amounts of detail. Large-scale processes therefore require other types of large-scale dynamic overview panels that can present the whole process in parallel.

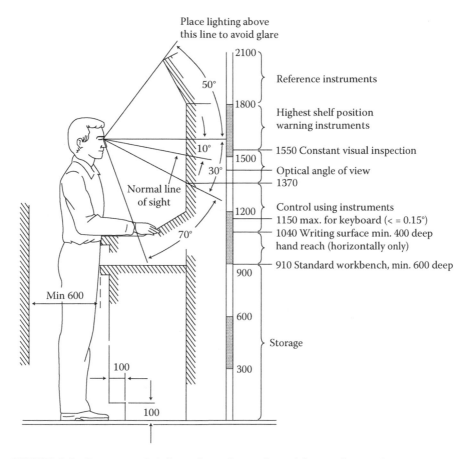

FIGURE 6.4 Recommended dimensions of control panel for standing work.

Work surfaces and controls must be positioned within easy and comfortable reach. This means having an area limited in height, width, and depth, within which the most important controls and work surfaces must be placed. Most types of control room work need plenty of space for carrying out the various tasks, apart from the control tasks as such. One needs, for example, writing surfaces, surfaces on which to place books and manuals, and surfaces for coffee and beverages. We look first at the heights of work surfaces and then at the surface area requirements.

The height of the work surface is of the greatest importance, whether the operator works standing or sitting. An incorrect work surface height results in the operator assuming a bad body posture, which in the short term is tiring and in the long term can cause permanent injury. With this in mind, we shall discuss first the positioning of the work surface itself. It should, of course, be pointed out that one should not view the work surface as an isolated unit, but the work surface, chair, and other equipment should be seen as a whole. Recommendations are given for both sitting

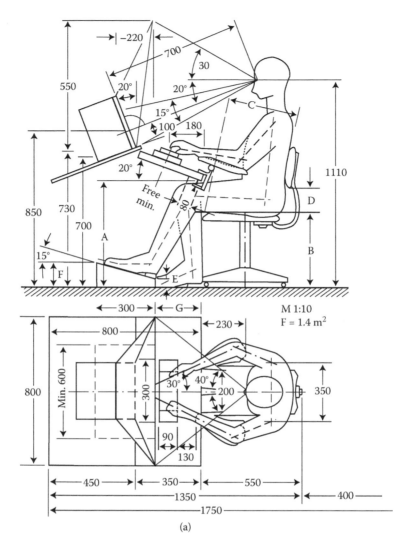

FIGURE 6.5(a)   Design of a CRT workstation.

and standing work. Sitting or standing workplaces are recommended for different types of jobs.

Sitting work positions are to be preferred to standing ones in the following cases:

1. In order to reduce fatigue. It is easier for an operator to work using his arms, and do heavier work with his legs, for a longer time when he is sitting rather than standing.
2. So that the operator can use both feet simultaneously and can develop greater force with his legs, use pedals faster, and operate several controls with one or both feet.

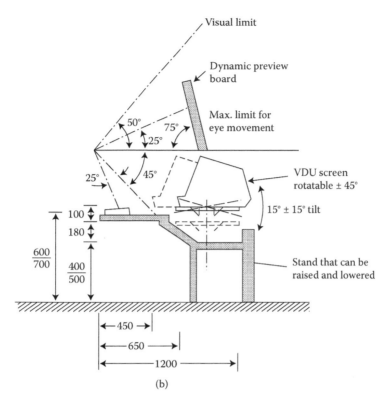

(b)

FIGURE 6.5(b)   An example of a process control workstation in a power station.

3. So that the operator can be protected from vibration while his arms and legs are free to use the controls.
4. When the operator's workplace is subject to movement. Some additional form of support for the body should be provided.

Standing work is to be preferred to sitting in the following cases (standing here means that the operator has the ability to move about):

1. Where greater mobility is required; by taking a step in a particular direction, the operator can reach considerably more controls.
2. Where the operator needs to carry out larger movements.
3. Where movements that require large force over large distances are carried out.

It is preferable for the work to be done both standing and sitting in turn, especially if the operator can choose when and if he or she wants to stand or sit. The workplace is then designed for standing work, and a high chair is provided. All work surfaces should be adjustable for height, so that they can be set at the height, which best suits each individual. If it is not possible to adjust the surface to the height required by each person, it is best to set up for the taller person rather than the shorter. A shorter person can then raise the chair, and at the same time use a footstool. A tall person, on the other hand, cannot become accustomed or adjust to working at too low a work height. The recommended heights of working surfaces for standing work are shown in Table 6.1. The recommended heights for sitting work are shown in Figure 6.6 and Table 6.2.

In the literature, several different heights are recommended for working surfaces. This is because sometimes the heights for men and women are differentiated. Alternatively, insufficient degree of adjustability may have been allowed for in the height of the surface or that a footstool has not been allowed for. In the 'Total' column in Table 6.2, measurements within which the working surface should be adjustable are indicated.

For sitting work, the work surface should be a few centimetres lower than the elbow height.

Figure 6.7 shows the areas of a horizontal working surface which can be reached with and without stretching. A workstation in a control room needs a free table surface for drawings, manuals, and handbooks. It must be possible to move the keyboard to increase the available area beside the CRT.

TABLE 6.1
Recommended Working Heights for Standing Work

Work Surface Height for Standing Work (cm)			
Work	Men	Women	Total
Precision work elbow support	102–112	107–117	102–117
Light work, work with small parts	92–102	97–107	92–107
Heavy work, lifting, pressing, etc.	77–92	82–97	77–97

Source: Grandjean (1988).

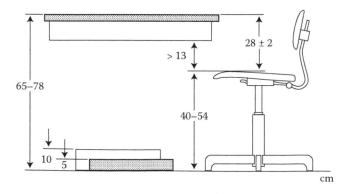

FIGURE 6.6   Workplace for reading and writing.

TABLE 6.2

Recommended Working Heights

Work Surface Height for Sitting Work (cm)			
Work	**Men**	**Women**	**Total**
Precision work at close distance	80–100	90–110	80–110
Reading and writing	70–74	74–78	70–78
Typing, handicrafts	65	68	65–70
Source: Grandjean (1988).			

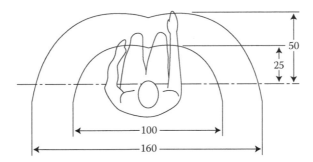

FIGURE 6.7 Vital measurements for all horizontal working surfaces.

6.5 SUPPORTING SURFACES

Standing for long periods causes fatigue, reduces performance, and is especially stressful for those workers with physical foot, leg, hip, or back problems. Such people are dependent on some form of supporting surface to relieve fatiguing and troublesome loading on the body. Supporting surfaces for sitting are the most common (see Figure 6.8). From a physiological standpoint, chairs should be designed so as not to cause pressure on the underside of the thigh. The weight of the body should be taken mainly on the ischial tuberosities, the bony points on either side of the base of the spine. This will avoid pressure on the underside of the thigh that results in a reduction in blood circulation in the legs, and numbs the large nerves that travel down the back of the legs. The resultant circulatory difficulties cause the legs and feet to 'go to sleep'. Pressure on the underside of the thigh can be avoided by making the chair sufficiently low, and reducing the degree of backwards slope of the seat. The chair should not, however, be so low that it causes pressure on the base of the spine. The height should be adjustable so that the chair gives suitable support to the tuberosities.

A chair should also be designed so that it gives suitable support to the back (particularly the small of the back). The support should not be so great that it hinders the movements of the arms and spine. The support should be concave in the horizontal planes and convex in the vertical plane. The back rest gives rise to a force that attempts to push the sitter out of his chair, and this may mean that the sitting person

A = Seat height   F = Backrest width
B = Seat width   G = Backrest height
C = Seat depth   H = Backrest centre height
D = Seat slope   I = Backrest radius
(E) = Seat concavity

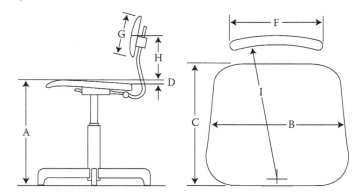

FIGURE 6.8   Vital measurements for a work chair. (E) *Note*: The concavity of the seat cannot be seen in the figure.

has to resist this force with the aid of the leg muscles; this in turn produces a static loading on the leg muscles. These static muscle forces can be prevented by means of a suitable backward sloping of the chair seat and sufficient friction on the seat.

A basic feature for comfortable sitting is that the chair must have a high coefficient of stability—that is, a high resistance to tipping over—which means that the base should be as wide as possible. If the floor is uneven, the chair should be adjustable, for example, by the use of screwed feet. Preferably, the floor should be level and evenly surfaced.

The supporting surfaces that we use may be categorised according to the function they fulfil. These functions can be divided into the following:

1. Resting surfaces (e.g., bed)
2. Looking/listening functions (e.g., TV armchair)
3. Working function (including looking, listening, speaking and working with objects; this may be occasional or full-time)
4. Special work (e.g., footstools)

6.5.1 WORK FUNCTION

The functions described in (2) and (3) involve looking, listening, and speaking while at the same time working with objects and tools. The function may either be full-time (such as a chair for typing work) or occasional (such as a visitor's chair or a conference chair). For the full-time function, it is important for the operator to be able to adjust every part of the chair so that it is suitable for just that one person. The chair does not need to be adjustable for the occasional function. There may even be advantages for *nonadjustable* chairs. For example, in some environments (such as a waiting room) difficulties may occur if every user were to make adjustments to the chairs. Instead, the chair should be designed so that it is suited to as many of the user population as possible. Regarding chair design, there are various seemingly conflicting philosophies. By studying all the arguments for and against these philosophies, it becomes clear that they are not in complete contradiction but to a large extent complement each other as they represent different types of sitting.

The four different philosophies are:

1. The 'classical sitting' style, first described systematically by Akerblom (1948) and later developed by Floyd and coworkers (1958, 1969), among others.
2. 'Active sitting', 'discovered' by Mandal (1982), and more recently marketed by the Scandinavian furniture company Hag.
3. 'Armchair sitting', proposed primarily by Grandjean (1988).
4. Culturally-specific sitting.

In this section we describe in detail the principles for these four types of sitting, together with some of their advantages and disadvantages.

1. *Classical sitting* is best suited to ordinary office work, where the work consists of many different tasks and where it has been possible to design the work surfaces to suit the work done on them. The sitting position starts from a relatively low (minimum height 35 cm) seat surface, in order to reduce pressure on the underside of the thigh, which leans slightly backwards to prevent the worker from sliding off. There is a small backrest that should provide support for the lower back and to help straighten it up. This position gives a limited range of movements and reach. When sitting correctly, the body is relatively relaxed and balance is achieved by voluntary action. The weight of the body is taken mainly on the ischial tuberosities (see Figure 6.9).
2. *Active sitting* is based on a higher seat tilted forwards, which straightens up the back 'naturally' (see Figure 6.10). The sitting position gives greater

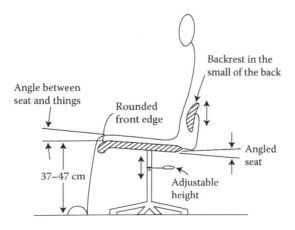

FIGURE 6.9 Classical sitting posture.

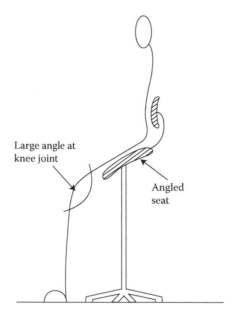

FIGURE 6.10   Active sitting posture.

freedom of movement and reach, achieves a better balance, and less room is needed for the legs. At the same time, the legs are subjected to a relatively high loading, as active sitting gives less support for the body. Recent studies show an increase in foot volume and foot complaints for this type of sitting. A correct posture for this type of sitting is adopted more or less automatically. The weight of the body is taken on the feet and the ischial tuberosities. It is well suited to those workplaces where long reach is required (due to a large work area) and/or where there is limited leg room. It is also suited to situations where there is a frequent change between sitting and standing. Combined with a sloping desktop it is also good for handwriting and ordinary reading tasks.

3. In *armchair sitting*, the body weight loading is distributed between a large part of the back, the ischial tuberosities, and the thighs. There is a large backrest and a large seat, both of which lean backwards. The sitting position is relatively low, but at the same time allows the best relaxation for the body and the best balance. It is best suited to looking/listening functions, for example, in conference rooms, control rooms, and for certain types of VDU work where the majority of the work is on the VDU with few other tasks (see Figure 6.11).

4. In *culturally-specific sitting*, there are many different types of traditions. Most of these can be related to cross-legged sitting directly on a floor or on any type of flat surface. For a person with sufficient flexibility in the legs to be able to sit cross-legged, this sitting position has many advantages. Automatically, the spine remains erect and in balance without tension in the

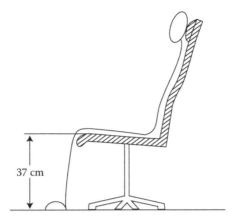

FIGURE 6.11   Armchair sitting posture.

back muscles. This is the obvious reason why this sitting position is used for mediation and in yoga. Preferably the sitting surface should not be too hard or too soft. Some people prefer to sit on a small cushion (a few centimetres high).

6.5.2   SPECIAL FUNCTIONS

In those cases where ordinary chairs cannot be used (for example, because the work surface is either too high or too low), special chairs can be used. For high work surfaces, load can be taken off the feet and legs by use of a 'shooting stick'. The following must be taken into account in the design of these:

1. They must be able to be raised and lowered.
2. The support on the floor must not be able to slide away.
3. The seat surface must have well-rounded edges.
4. The seat surface must be padded.
5. The seat material must have good friction.

Figure 6.12 shows a special chair which can be used at both high and low workplaces. It is designed so that it can be easily moved between different workplaces. It can be raised from 240 mm to 660 mm above the floor. The seat depth is also adjustable and can be turned through ±30 degrees. The seat surface is so designed that the operator has support for the ischial tuberosities, while at the same time providing relatively good freedom of movement for the legs. The chair is a combination of an ordinary chair and a standing support chair/stool.

Chairs for all types of functions are needed in the control room. For normal monitoring work where no control or recording function is carried out, the looking/listening chair is perhaps the best. For more intensive recording and keyboard work and writing, the chair with the working function is the best. When working for

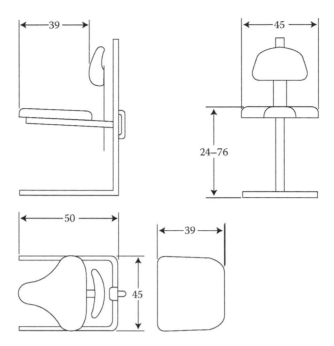

FIGURE 6.12   An example of a 'standing–sitting' chair adjustable over a wide range.

longer periods at a panel, where normally only occasional readings have to be taken, there may be a need for special chairs of the support type. There is no one combination chair that fulfils all these three functions. However, attempts have been made to produce chairs that satisfactorily fulfil working and certain resting functions (see, for example, Figure 6.13).

6.6   EXAMPLES OF CONTROL ROOM DESIGN IN A PROCESS INDUSTRY

Figure 6.14 shows an example of a layout for a control room. A workplace is placed centrally with access to a video display terminal (VDT), keyboard, and special controls. There is also space on the central workplace for writing and for various reference books. The process, which is assumed to be large, is displayed on the overview panel, mainly in the form of static information, but with the insertion of certain dynamic information such as the settings of the more important controls, critical readings, or deviations from the set values. The overview board is designed so that all parts of it are at more or less the same distance from the operator. There is also a more traditional office workplace in the control room, where various special tasks can be carried out during the daytime.

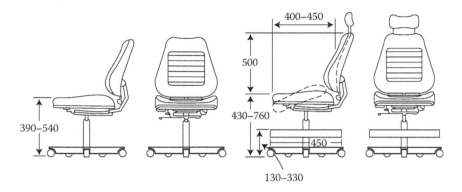

FIGURE 6.13 Chair for both working and listening/looking, together with certain resting functions.

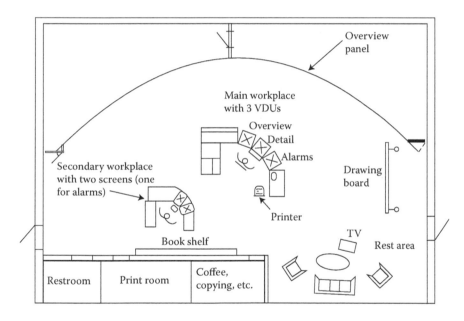

FIGURE 6.14 Example of a process control room design.

In a control room where several people are working and there is an overview panel, it is important that this is positioned so that everyone can see it.

Figure 6.15 shows a more detailed design of a terminal workplace and for many ordinary office workplaces. Here, it is important that the workplace really is flexible with easily adjustable heights and distances of its various parts. If possible, it should also be possible to angle the keyboard and screen as shown in Figure 6.5b.

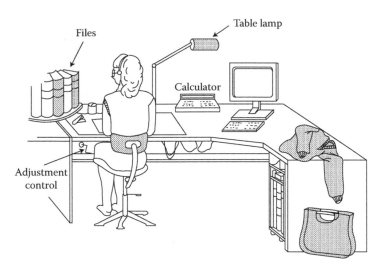

FIGURE 6.15   Example of an ordinary terminal workplace.

REFERENCES AND FURTHER READING

Aberblom, B. (1948). *Standing and Sitting Posture.* Stockholm: Nordista Bokhandelns, Förlag.

Anshel, Jeffrey. (2005). *Visual Ergonomics Handbook.* Boca Raton, FL: CRC Press.

Ayers, Phyllis A., and Kleiner, Brian H. (2002). New Developments Concerning Managing Human Factors for Safety. *Managerial Law* 44, 1/2, 112–20.

Desai, Anoop, and Mital, Anil. (2005). Incorporating Work Factors in Design for Disassembly of Product Design. *Journal of Manufacturing Technology Management* 16, 7/8, 712–32.

Doggette, William T. (1995). Office Ergonomics. *Benefits Quarterly* 11, 4, 21–27.

Floyd, W.F., and Roberts, D.F. (1958). Anatomical and Physiological Principles in Chair and Table Design. *Ergonomics* 2, 1–16.

Floyd, W.F., and Ward, J. (1969). The Sitting Posture. *Ergonomics* 12, 132–39.

Grandjean, E. (1988). *Fitting the Task to the Man* (4th ed.). London: Taylor & Francis.

Mandal, A.C. (1982). *Den Sittande Människan.* Stockholm, Sweden: Liber Förlag.

Meshkati, Najmedin. (2003). Control Rooms' Design in Industrial Facilities. *Human Factors and Ergonomics in Manufacturing* 13, 4 (Summer), 269–77.

Noyes, Jan, and Bransby, Matthew (eds.). (2001). *People in Control: Human Factors in Control Room Design.* London: IEE Press.

Valenti, Michael. (1995). A New Generation of Nuclear Reactors. *Mechanical Engineering* 117, 4 (April), 70–75.

Vink, Peter (ed.). (2004). *Comfort and Design: Principles and Good Practice.* Boca Raton, FL: CRC Press.

7 Environmental Factors in the Control Room

Andy Nicholl

CONTENTS

7.1 INTRODUCTION

Environmental factors can be categorised as either physical or chemical in nature. Chemical factors usually include dust, gases, vapours, smoke, and various different chemicals. This type of problem hopefully has little relevance to most control room situations. However, there are a large number of problems that belong to the physical category, such as thermal climate (warmth, cold, humidity, and draught), lighting, and acoustic climates, which are of particular relevance in the control room situation. Problems associated with the thermal climate are underestimated in most types of control rooms. It is important when dimensioning air-conditioning systems to take account of the heat given off by the different pieces of equipment in the control room, as these can cause a considerable addition to room heat. This chapter briefly describes various requirements for thermal comfort, lighting, and acoustic climate with relevant examples.

Noise arising in the control room can often be distracting, particularly during speech or telephone communication. This noise can arise from printers and fans. Particular tones may also be given off by different pieces of electronic apparatus, and the air-conditioning system can give rise to disturbing levels of noise. Vibration of the floor may occur in control rooms attached to a number of industrial processes. Whole-body vibration of this type may lead to fatigue in the workforce exposed to it.

The most important environmental factor in the control room is usually the lighting. Both in the use of visual display unit (VDU) screens and in more traditional control rooms, the demands on the light fittings are very high. Where VDUs are in use it is important to ensure, among other things, that there are no distracting reflections on the screen or keyboard. When using a modern VDU screen with diagrams and small characters, the light level should be a maximum of 200 to 300 lux, depending on the type of visual task occurring in the work. A suitable type of supplementary workplace light is often required.

The physical environment should be designed in such a way that the risks of injury and health problems are avoided. On the other hand, the environment should be designed so that it is comfortable and causes as little trouble and disturbance or distraction as possible for the user. The environmental conditions should be designed so that they affect people's working ability as little as possible.

7.2   THERMAL CLIMATE

The human organism is so designed that it strives to maintain a constant internal temperature even during high levels of physical workload. For low physical work-loads this temperature is 37°C, with certain upward variations in daytime and down-ward variations at night. If the environmental conditions are such that the body's ability to maintain its temperature at the required level is exceeded, a lack of comfort is experienced, and if the stress is sufficiently great there may be detrimental effects on health. But problems or lack of comfort can also occur without any change in the body's core temperature. Local cooling or warming can be experienced as uncomfortable and help to reduce the level of a person's mental performance. Local cooling may arise, for example, from too great a temperature difference between feet and head or between the two sides of the body.

Thermal comfort is a condition where people experience no discomfort from the thermal climate, that is, they do not know whether they want it warmer or cooler. People are different in their individual preference for climate, and it is therefore impossible to create a climate that satisfies everyone whenever there is more than one person in a room.

7.2.1   WHAT DETERMINES THE CLIMATE?

Human experience of thermal climate is determined by a number of factors in the physical environment. It is affected by the ordinary air temperature (T_a), the temperature of the surrounding surfaces (T_g: black globe temperature, related to radiant temperature), air humidity (RH, %), and air movement (V, ms^{-1}). Other important factors include activity level and clothing. There are also small differences due to gender, race, and age, but these factors are of lesser importance. The metabolic processes of the body produce warmth; increasing the workload increases heat production. Heat production as a function of activity, clothing, and air speed is given in Table 7.1. Control room work with a low loading gives a heat production in the body of 100 to 130 W, and 160 to 170 W at high loads.

7.2.2 TEMPERATURE INDICES

In order to determine the loading on people from the thermal climate, various types of temperature indices can be used. Different climatic indices are used depending on whether one is determining the dangers to health or the comfort levels. The index most commonly used to determine the effects on health is the wet bulb globe temperature (WBGT) index:

$$WBGT = 0.3\ T_g + 0.7\ T_w \qquad (7.1)$$

where T_g is the globe temperature and T_w is the wet bulb temperature.

Globe temperature is measured using either an electronic or a common mercury-in-glass thermometer mounted inside a black globe 150 mm in diameter, and it gives an estimation of the radiant temperature; it is affected by the air temperature, the temperatures of the surrounding surfaces, and the air speed. T_w is the wet bulb temperature, which is measured with an ordinary thermometer whose bulb is surrounded by a wetted 'sock'. This should give the 'natural' wet bulb temperature.

Where a fan is used to force air past the temperature-sensitive wet bulb, the 'psychrometric' wet bulb temperature is obtained. The natural wet bulb temperature represents the air temperature corrected for the relative humidity in the room, and with a certain effect from the air movement where this is below about 2 ms^{-1}. Lower relative humidities and higher air movements give a lower temperature on the natural wet bulb thermometer compared with the ordinary air temperature. Figure 7.1 shows the acceptable WBGT values for different time periods and levels of activity.

The area of particular interest in the control room is the 'comfort region', and a special comfort index may be used to determine this. Many comfort indices have been produced. The one with the best foundation is that produced by P.O. Fanger (1970), called the Fanger index.

7.2.3 DETERMINATION OF THE COMFORT CLIMATE

Figure 7.2 and Figure 7.3 show summaries of the comfort criteria according to Fanger (1970). The complete calculations and evaluations as produced by Fanger are relatively complicated. Here it need only be said that if Fanger's comfort criteria are wholly fulfilled, an estimated minimum of 5% of people will still be displeased with the climate (for some people, it will always be too warm or too cold). In other words, this is the lowest percentage of dissatisfaction that can be reached. One criterion for the climate conditions to be fulfilled is that the climate must be consistent throughout the whole room, with no variation in different areas or between the floor level

TABLE 7.1

Heat Production from an Average Person as a Function of Activity, Clothing, and Air Speed

Activity	Clothing	Relative Air Speed	Comfort Temp (MRT = Air T)	Air T drop when MRT rises 1°C	Heat Produced by Average Person (MRT = air temp)		Water Vapour Production		Total
					Convection	Radiation			
		m/s	°C	°C	W	W	g/h	W	W
Sedentary	None	<0.1	28.8	1.00	36	38			
		0.3	30.1	0.50	47	29	40	27	
		0.5	30.7	0.45	51	24			
		1.0	31.4	0.35	57	20			
	0.5 clo[b]	<0.1	26.2	0.95	36	37			
		0.3	27.4	0.60	47	28	42	28	
		0.5	27.9	0.45	50	23			
		1.0	28.6	0.30	55	17			
	1.0 clo	<0.1	23.3	0.95	36	35			
		0.3	24.5	0.60	45	27	44	30	102
		0.5	25.0	0.45	50	22			
		1.0	25.5	0.30	55	17			
	1.5 clo	<0.1	20.7	0.90	36	34			
		0.3	21.8	0.55	45	26	46	31	
		0.5	22.3	0.40	50	21			
		1.0	22.8	0.30	55	16			

Activity / Clothing								
Medium activity level								
None	<0.1	24.4	1.00	59	65			
	0.3	26.2	0.65	76	51	115	77	
	0.5	27.1	0.50	83	44			
	1.0	28.2	0.35	93	35			
0.5 clo	<0.1	19.9	0.95	60	63			
	0.3	21.6	0.60	76	48	120	80	
	0.5	22.4	0.50	83	41			
	1.0	23.3	0.35	92	33			
0.1 clo	<0.1	15.3	0.90	60	59			204
	0.3	16.9	0.55	76	45	123	83	
	0.5	17.7	0.45	83	36			
	1.0	18.6	0.30	91	30			
1.5 clo	<0.1	10.9	0.85	62	57			
	0.3	12.5	0.50	77	43	126	84	
	0.5	13.2	0.40	83	36			
	1.0	14.0	0.30	91	29			
High activity level								
None	0.3	22.1	0.60	107	67			
	0.5	23.4	0.50	117	60	192	129	
	1.0	24.9	0.35	130	48			
	0.3	9.3	0.60	110	59			
0.5 clo	0.5	16.8	0.45	119	55			
	1.0	18.2	0.35	130	44	198	129	
	0.3	9.3	0.60	110	59			

TABLE 7.1 (continued)

Heat Production from an Average Person as a Function of Activity, Clothing, and Air Speed

Activity	Clothing	Relative Air Speed	Comfort Temp (MRT[a] = Air T)	Air T drop when MRT rises 1°C	Heat Produced by Average person (MRT = air temp)		
					Convection	Radiation	Water Vapour Production
		1.0	11.7	0.30	131	40	
		0.3	3.2	0.45	113	56	
	1.5 clo	0.5	4.2	0.35	122	47	205
		1.0	5.4	0.25	131	37	135

[a]MRT, mean radiant temperature.

[b]The amount of clothing is denoted by a "clo" value, which is a measure of the thermal resistance of the clothing. 1 clo = 0.154 m.K/W, which corresponds to "normal" clothing.

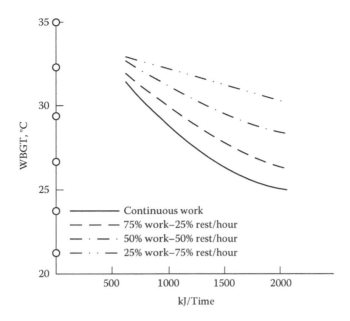

FIGURE 7.1   Recommended limit values for heat exposure with workload.

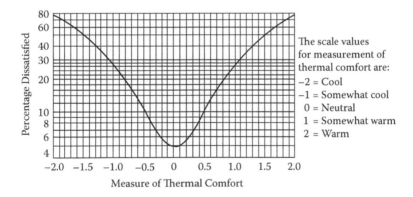

FIGURE 7.2   Proportion of people in any one environment who will be dissatisfied with their thermal comfort.

and the ceiling level, for example. Neither should there be any cold surfaces in the room which could cause cooling or 'cold radiation'—that is, cold surfaces towards which the body emits radiant heat but receives little in return—resulting in a local cooling of the body. This can include window areas.

   It is most common to find sedentary work in the control room and for operators to be lightly clothed. In this case the relative humidity should be around 50%, the maximum air speed should be 0.1 ms^{-1}, and the air temperature should be 26°C. If it is

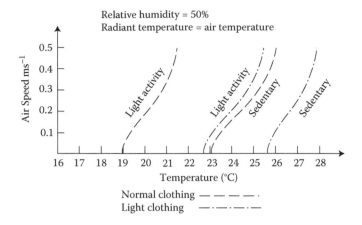

FIGURE 7.3   Effect of air speed on optimum comfort temperature.

felt from an economic point of view that this is too high a temperature, the operators would have to be asked to wear somewhat warmer clothing such as long trousers and a jacket or pullover. With this type of clothing being worn, the conditions should be 50% relative humidity and 0.1 ms⁻¹ air movement with an air temperature of 23°C.

In control rooms where there is a certain amount of physical activity, such as walking between the various control panels and working mainly in a standing position, the temperature (with normal clothing) can be dropped to 19 to 20°C and air speeds up to 0.2 ms⁻¹ can be accepted. The air speeds in the room will increase in any case due to the movement of people in the room.

In many control rooms the work is more sedentary at night than in the daytime and it may therefore be best to have slightly higher temperatures at night, 21 to 22°C compared to 19 to 20°C during the day.

As different people have varying requirements regarding what is an acceptable climate, the operators should be able to regulate the climate. The air humidity should be kept between 40 and 60%. If the humidity is low, the air feels cooler, so higher air temperatures are needed. Low humidities can cause a drying-out of the mucous membranes of the nose and throat, which in turn can lead to an increased risk of throat and chest infections.

7.3   LIGHTING CONDITIONS

It is outside the scope of this book to describe in any great detail the basic technical concepts behind lighting principles. It is sufficient here to state that the unit used for the intensity of illumination at a particular surface—that is, the quantity of light falling on a surface from different light sources—is the lux (lumen per square metre or lumen/m⁻²).

An ordinary candle at a distance of 1 metre gives an intensity of about 1 lux; at 2 m distance, 1/4 lux; at 3 m distance, 1/9 lux; and so forth. In other words, the light intensity is inversely proportional to the square of the distance. Sunlight can give light intensities of up to 100,000 lux. The luminance is the 'lightness' of a surface. This may either be a surface that lights up itself, for example, a VDU screen, a light, or a surface that reflects light from other light sources. The usual unit for luminance is the candela per square metre (cd/m^2). Table 7.2 summarises a number of technical terms, units, and relationships.

There are three main areas of human requirements that must be considered when designing lighting installations:

1. Injuries (safety)
2. Performance effects
3. Comfort

These requirements are usually considered in other types of ergonomic dimensioning work. Both quantitative and qualitative aspects of these three demands will be examined in relation to the design of lighting installations.

7.3.1 LIGHTING REQUIREMENTS IN THE ROOM

The lighting intensity in a room must not be so high that there is any risk of injury. Considerable problems will arise if the light intensity is high and the light fittings are of an unsuitable design that gives rise to glare. Too low a lighting level also reduces performance considerably. Research has shown that visual performance varies according to different lighting variables (Hopkinson and Collins, 1970). The variables that affect visual performance are:

1. The luminance in the room and especially around the visual object.
2. The critical size of the visual object, i.e., the smallest element that must be distinguished in order for the whole object to be distinguished.
3. The contrast of the visual object, i.e., the relationship between the dark and the light parts of the visual object that have to be distinguished in order for the visual object itself to be distinguished.
4. The viewing time available.

Figure 7.4 summarises the way in which visual performance varies with the different variables.

A suitable illumination level in a control room is about 1000 lux or higher at night-time to keep the operators' arousal at a high level. If VDUs are in use, however, the illumination level should be restricted to 300 lux. The luminance distribution in the room has to be comfortable. This is determined by the reflection factors on the surrounding surfaces. The reflection factors are the proportion of light reflected or retransmitted from a surface, varying from 0.0 for totally black to 1.0 for total light reflection. The reflection factors of the room surfaces should be about 0.2 to 0.4 on floors, 0.4 to 0.6 on walls, and about 0.4 to 0.8 on the ceiling, with the lower values being suitable if there is risk of reflections occurring in VDU screens.

TABLE 7.2

Definitions and Conversion Factors for Lighting Units

Standard Units, Symbols, and Defining Equations

Quantity	Symbol	Define Equation	Unit	Symbolic Abbreviation
Luminous energy (quantity of light)	Q	$Q = \int \Phi \alpha$	lumen-hour lumen-second	lm.h lm.s
Luminous flux	Φ	$\Phi = dQ/dt$	lumen	lm
Luminous excitance	M	$M = d\Phi/dA$	lumen per square metre	lm/k²
Luminance	E	$E = d\Phi/dA$	lux	lx
Luminous intensity (candle power)	I	$I = d\Phi/d\Omega$ (Ω = solid angle through which flux from point source is radiated)	candela (lumen per steradian)	cd
Luminance	L	$L = dI/dA \cos\theta$ (θ = angle between line of sight and normal to surface considered)	candela per square metre	cd/m²
Luminous efficacy	K	$K = \Phi_v / \Phi_s$	lumen per watt	lm/W
Scalar luminance	E_∞	$E_\infty = \int E/4 x d\Omega$	lux	lx
Light exposure	H	$H = dQ/dA$	lux-second	lx.s
Absorbance	α	$\alpha = d\Phi_s/d\Phi_a^{\,\circ}$	(numerical ratio)	—
Reflectance	p	$p = d\Phi_v/d\Phi_a^{\,\circ}$	(numerical ratio)	—
Transmittance	γ	$\gamma = d\Phi_z/d\Phi_a^{\,\circ}$	(numerical ratio)	—
Conversion Factors	Candelas per sq metre	stilb	footlambert	
Luminance Units	(cd/m²)	(cd/m²)	(fl)	

Multiply by

Candelas per sq metre (cd/m²)	1	10 000	3.43
Stilb (cd/m²)	1×10^2	1	3.4×10^1
footlambert (fl)	0.2919	2919	1
footcandle (lm/ft²)	0.0929	0.0929	10.764

Notes: 1 footcandle (fc) = 10.764 lux; 1 lux = 0.0929 fc.

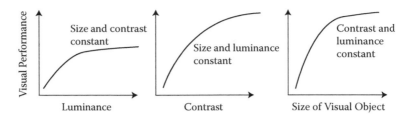

FIGURE 7.4 Variables affecting visual ability. (From research by Weston, 1962. With permission.)

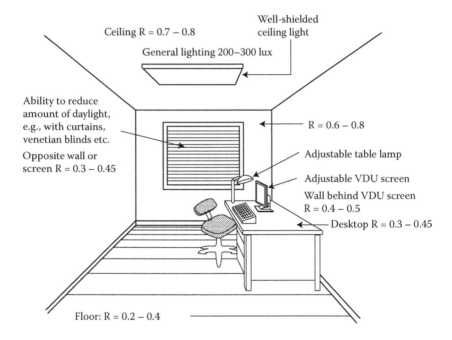

FIGURE 7.5 Acceptable reflection factors when using VDUs.

The lighting levels on the working surfaces and control panel should be greater than those in the room in general. These surfaces should also have a rather higher reflection factor than the room in general in order for the luminance to be higher on these surfaces; that is, they should stand out as being brighter than the overall room luminance. However, it is important for panels and worktables not to be shiny, otherwise reflections would result. Figure 7.5 gives suggestions for acceptable reflection factors in places where VDUs are used.

The design of the light fittings in the control room is very important. In most types of control rooms, particularly those with VDUs, it is very important that the

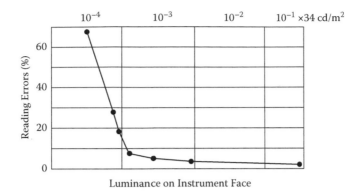

FIGURE 7.6 Reading errors due to luminance.

lights do not cause reflections in the object being viewed. It may often be best to use reflective (for example, 'Paracube') shading under fluorescent strip lights, as these direct the light vertically downwards and give a relatively dark surface when seen from the side.

7.3.2 LIGHTING OF DIAL AND METRE TYPES OF INSTRUMENTS

In certain situations a minimal level of lighting is required on older types of dial and meter instruments because, among other things, the eyes have to be dark-adapted while using them. Figure 7.6 shows the possible reading errors due to poor luminance on the instrument face. They all should have the same level of lighting. A ratio of 7:1 can be allowed between the most strongly and the most dimly lit instruments.

Apart from instrument lighting, all direct light towards the operator's eyes must be avoided. Table 7.3 gives recommended values for instrument lighting.

There are two principles in the lighting of instruments, direct and indirect, and these will be discussed further.

7.3.3 DIRECT LIGHTING

This method of lighting is independent of the instrument itself. The lighting can be positioned over the instrument and in this way reduce reflections from it. The reflected light is directed downwards. The advantages of this method of lighting are:

1. Lighting is even.
2. Controls can also be lighted.
3. The space between the instruments is lit.
4. Broken bulbs can easily be changed.

TABLE 7.3

Recommendations of Illumination of Instruments

Use	Recommended Lighting Technique	Luminance of Markings (cd/m)	Adjustability
Meter reading, dark adaptation necessary	Red directed light, direct or indirect, operator's choice	0.07–0.2	Continuous
Meter reading, dark adaptation not necessary but desirable	Red or white light with low colour temp., direct or indirect, operator's choice	0.07–3.4	Continuous
Meter reading, dark adaptation not necessary	White directed light	3.69	Fixed or continuous
Reading of text on panel, dark adaptation not necessary	Red built-in light or directed light or both, operator's choice	0.07–0.3	Continuous
Reading of text on control panel, dark adaptation not necessary	White directed light	3–69	Fixed or continuous
Operator may be exposed to sudden bright light sources	White directed light	34–69	Fixed
Very great heights, minimal daylight due to design of cockpit	White directed light	34–69	Fixed
Diagrams and chart reading, dark adaptation necessary	Red or white directed light, operator's choice	0.3–3.4	Continuous
Diagrams and chart reading, dark adaptation not necessary	White directed light	17–69	Fixed or continuous

The disadvantages with this method of lighting are:

1. It may be difficult to position the lamp so that it does not block the area of view.
2. Shadows may be produced by the edge or pointer of the instruments.

7.3.4   INDIRECT LIGHTING

With indirect (in-built) lighting, the lighting is built into the instrument. The advantages are:

1. The light does not spread to other areas.
2. The lighting can be tailor-made for each individual instrument.
3. The light fitting or lamp does not block anything in the workplace.

The disadvantages are:

1. The area between the instruments is not illuminated, which may make an instrument look as though it is 'floating in midair'.
2. It may be difficult to achieve even lighting over the face of the whole instrument.

The most common in-built lighting systems are:

Indirect lighting—The lighting is placed behind the instrument face, which is made of a material which lets sufficient light through.

Edge lighting—This lighting technique works by manufacturing the instrument out of a material, which normally permits light to pass through. The instrument is then painted completely black, apart from the figures, which are to be read. One advantage for this technique is that different colours can be used for different instruments.

Self-illuminating (fluorescent) markings—Here two techniques can be used: fluorescent or electro-fluorescent. Electro-fluorescent markings have the advantage that they are only illuminated when current is applied. With these techniques, only the markings and the numbers/letters on the instrument are seen. The advantages of these techniques are:
1. Even lighting and colour of the markings
2. Ability to change the colours

7.3.5   LIGHTING REQUIREMENTS WHERE VISUAL DISPLAY UNITS (VDUs) ARE USED

Serious visual problems can easily arise in work on VDUs, and this can cause a lack of visual comfort, subjective eye troubles, and reduced levels of work performance. The problems can be reduced by having a suitable distribution of luminances in the room and at the workplace, examples of which are shown in Figure 7.5 and

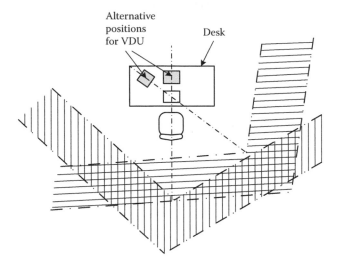

FIGURE 7.7   Ceiling lights must not be placed in the area marked with a vertical stripe for VDUs directly in front of the operator, or in the horizontally striped area for VDUs placed diagonally to the left of the operator.

Figure 7.7. When working with VDU screens with light text on a dark background, the lighting levels in the room must be kept relatively low. However, this must not be so low that operators cannot do other visual tasks (for example, reading printed or handwritten text). A suitable compromise is a general lighting level of 200 to 300 lux with table lighting available which can give a local light level of about 600 lux for reading difficult handwritten text or smaller printed text.

Reflections on the keyboard must be avoided. This can be achieved by leaving a light-free area over the keyboard. The size of this area is shown in Figure 7.7 and depends on the height of the light above the table.

The ceiling lights must also be placed in such a way that they do not give rise to reflections in the VDU screen. This can best be done by placing the light close enough to the screen so that the reflection occurs below eye height. Practical experience has shown that there is a limited 'wedge-shaped' area (see Figure 7.7) within which the lights should be placed. If the lights are installed behind this limit line, there is a risk of reflections occurring. This, however, reduces with distance, and the risk of reflection is even less if the lamps are shaded (for example, with a Paracube) so that they appear dark if viewed from the side. Many lamps suitable for use in connection with VDU work have a negligible luminance when viewed from an angle greater than 20° from the light axis.

If the limitations regarding reflections on the keyboard and in the screen are combined with suitable angles for the light fittings, the result is as shown in Figure 7.7. Lights should not be positioned within the hatched areas.

In the terminal control room, one can either choose lights that give a mainly downwards-directed light, together with sharp edge lines, or ones with 'softer' edge lines. If the light in the first case is viewed from a sufficiently acute angle, it will be seen as being relatively dark and the risk of reflection is relatively small. The luminance should be a maximum of 100 cd/m^2 at a viewing angle of up to 20°. The lighting efficiency of such a lamp is relatively good, but the limiting edge lines are very sharp and a small mispositioning of the screen in relation to the light fitting can cause severe reflections. If it is known that more flexible furnishing is necessary, it is better to choose a light fitting that gives softer shadows and a certain degree of indirect lighting. Table lighting should be arranged so that it does not cause reflections on the screen. This means that the table lamp must not be able to be turned in such a way that the light source can be seen. The light should instead be able to be adjusted by means of movable reflectors placed within the shade. These reflectors can then be used to direct the light towards different areas of the table.

It may be necessary to obtain special spectacles for those with visual deficiencies in order to provide them with good visual conditions. Ordinary spectacles are often designed to allow for reading at a normal distance, ca. 50 cm. Spectacles therefore need to be specially ground for reading a VDU screen at a closer distance. If there are also overview panels that have to be read at several metres' distance, a third ground area in the spectacles is required for the trifocal or multifocal glasses. When ordering spectacles of this type, an optician must be contacted who has a special interest in this type of problem. Information on the type and nature of the work should be provided in order for the correct grinding to be carried out.

7.4   ACOUSTIC CLIMATE

The levels of noise present in control rooms are not generally high enough to cause hearing damage. Nonetheless, considerable disturbance can be caused by noise. Noise is defined as an undesirable sound. It is produced as a succession of pressure waves in the air from sources such as vibrating surfaces (e.g., loudspeaker cones, equipment panels, or a person's vocal cords) or air turbulence caused by, for example, fans or compressed air outlets. The noise level or 'sound pressure level' is usually measured in decibels (dB).

The full frequency range of human hearing lies between 20 Hertz (Hz) and 20,000 Hz. As the human ear is less sensitive to sounds at low frequencies and very high frequencies, a special weighting is applied to sound measured at different frequencies so that their effect can be compared directly. Known as the A-weighting, this mimics both the sensitivity of the ear to sounds at different frequencies and also the likelihood of damage occurring to the ear at different frequencies. The A-weighted decibel is written dBA, although the older notation dB(A) is still seen.

Following a European Directive published in 2002, all countries in the EU now have noise regulations that limit the daily average noise levels to which people may be exposed at work. The Lower Exposure Action Value (LEAV) is an average over the day of 80 dBA, at which point a noise risk assessment should be carried out. The Upper Exposure Action Value (UEAV) is set at 85 dBA, and this is taken as the point at which the risk of hearing damage starts to rise (with consequent implications for health surveillance). The regulations set a daily limit value (ELV) of 87 dBA. Each of these criterion values has an associated peak limit level, but sudden peaks of noise should not normally be a problem in control rooms.

7.4.1   Noise Sources and Measurement

Noise in control rooms can arise from sources within the room, such as fans and impact (dot matrix) printers. Fans used to provide cooling to electronic and computerised equipment are often a significant source of control room noise. Impact printers have now largely been replaced by inkjet or laser printers, which are generally very quiet in operation. External noise sources include air-conditioning systems, where the refrigeration pumps, circulating fan, and turbulence in the air emerging from the vents can all transmit noise and raise the noise level in the control room. Air turbulence tends to produce higher-frequency noise. Control rooms are often attached to the structure of the building in which the industrial process being controlled takes place. Where the process produces vibration, this can be transmitted through the walls and to the equipment panels in the control room, and this in turn creates noise in the room. This type of noise tends to be in the low-frequency range.

In an open space, the noise level reduces by 6 dB for every doubling of distance from the source in accordance with the inverse square law. In an enclosed space such as a control room, however, with sound-reflective surfaces, the level drops only while one is close to the source. At a greater distance the noise level becomes constant due to the reflections in the room. The noise level will only drop by between 3 and 5 dB, even at the corner of the room furthest from the sound source. The noise level in a control room can be estimated by the operators using a simple rule of thumb as shown in Table 7.4.

Control room operators tend to spend long periods in the control room environment. If the noise level in the control room is above 85 dBA, then the operator's daily average exposure is also likely to be above 85 dBA and action would then need to be taken to reduce the exposure of the operator. The best solution to this is to investigate the origin of the noise and to reduce it at source. This then protects everyone in the room, whether operator or visitor. The use of personal ear protection is not usually recommended in control rooms as it reduces the levels of all sounds, not just the background noise, and it can make speech much more difficult to hear. There may be situations, however, where the wearing of hearing protection may be beneficial.

TABLE 7.4
Noise in the Workplace Environment

Test	Probable Noise Level	A Risk Assessment Will Be Needed If The Noise Is Like This for More Than:
The noise is intrusive but normal conversation is possible	80 dBA	6 hours per day
You have to shout to talk to someone 2 metres away	85 dBA	2 hours per day
You have to shout to talk to someone 1 metre away	90 dBA	45 minutes per day

Source: Hse (2005).

7.4.2 Noise and Communication

The most important problem with control room noise is that the noise disturbs communication between people and also makes it difficult for the operators to hear the various signals from both the control equipment and directly from the process. In order for speech to be easily understood, its level must be considerably greater than the background level in the room, as some individual speech sounds are much lower than the average level of the speech. If the average speech level is approximately the same as the background level in the control room, speech would still be mostly intelligible. If it lies at about 5 dB below the background level, it will only be partially intelligible even with concentrated listening.

The speech level of operators in a control room will vary widely depending on its strength, but a common value lies between 60 and 65 dBA at a distance of 1 metre. This means that for speech at a normal conversational level to be heard properly the background noise level should not exceed around 55 to 60 dBA. One measure of the interfering effect of noise on speech is the Speech Interference Level (SIL) which was defined by Beranek as the arithmetic mean of the sound pressure level in the octaves: 600 to 1200, 1200 to 2400, and 2400 to 4800 Hz (Beranek, 1954). In line with modern sound level metres, the SIL is now calculated as the arithmetic mean of the noise levels (in dBA) in the octave bands centred around 1 kHz, 2 kHz, and 4 kHz. This set of graphs is shown in Figure 7.8.

7.4.3 Noise and Masking

Noise can mask the various signals that it may be necessary for the operator to hear as it raises the threshold of audibility of such signals. The effect of masking is very complex, and the particular masking effect caused by a noise depends on its level and spectral composition. Where a noise consists mainly of a single tone, the greatest degree of masking will occur at a frequency one octave higher. It can be seen from Figure 7.9, for example, that a tone or a noise with a main frequency of 200 Hz and a level of 60 dB will completely mask a signal of 1000 Hz at 25 dB (everything under the continuous line is masked).

On the other hand, if the level of the masking noise at 200 Hz is dropped to 40 dB, the signal will be heard, as only sounds under the dashed line will be masked. Figure 7.9 also shows the masking effect of 60 and 30 dB tones at 3500 Hz. Broadband noise (for example, 'white' or 'pink' noise) will cause masking over most of the auditory spectrum, although higher frequency noise from fans and air-conditioning will affect the intelligibility of speech more than lower frequency noise.

Figure 7.9 shows that the masking effect is greater for frequencies higher than the frequency of the masking noise itself. Different types of noise are associated with different degrees of disturbance:

More Disturbing	**Less Disturbing**
High frequencies	Low frequencies
High noise levels	Low noise levels
Intermittent noise	Continuous noise
Unexpected noise	Predictable noise
Unexplained or unreasonable noise	Noise that appears reasonable

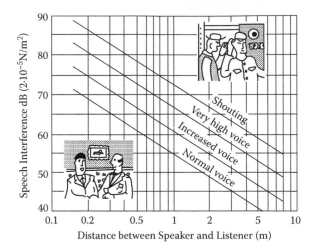

FIGURE 7.8   Relations between voice level, distance, and speech interference level (SIL). SIL is defined as the arithmetic mean of the sound pressure level in the octaves: 600 to 1200, 1200 to 2400, and 2400 to 4800 Hz (according to Beranek, 1954). The curves show the distance between the speaker and the hearer (*x* axis) and the maximum background noise levels in which speech is intelligible for communication using a normal, raised, very loud, and shouting voice.

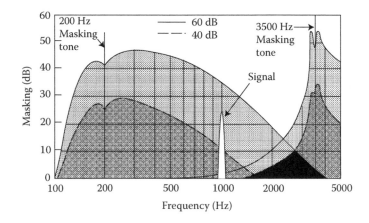

FIGURE 7.9   Masking produced by pure tones. (VanCott and Kinkade, 1972. With permission.)

Noise is known to cause increased levels of stress in the human body, but although the effect of noise on accident frequency has been studied, no clear-cut results have emerged. There is a tendency, however, for accidents and mistakes to occur more frequently in noisy workplaces, especially where a certain degree of concentration (on machines, screens, and so forth) is required. Noise, especially intermittent or high frequency noise or noise at a high level, will cause more errors in the operator's work.

7.4.4 NOISE REDUCTION

To maintain efficiency and for good communication, the noise levels in control rooms should be kept as low as possible. There are three main points at which noise can be reduced before it is perceived by the ear: the source, the pathway, and the receiver:

$$Source \rightarrow Pathway \rightarrow Receiver$$

The following are three ways in which noise levels can be reduced:

1. Reduce the noise level at the source: i.e., take action at the source.
2. Prevent the noise spreading from the source: i.e., take action at the pathway.
3. Give the person affected some form of hearing protector: i.e., take action with the receiver.

Hearing protectors should not be used in the control room. The work is such that the use of hearing protectors will cause communication problems and could also disturb the work. An attempt should be made to prevent noise from being created. The following methods may be used to reduce the noise generated:

1. The forces that cause the vibration of the different machine or equipment parts should be localised as much as possible, reduced, or, if possible, eliminated. Rotating parts such as fan blades should be well balanced, and fans should be mounted in heavy casings with flexible mountings to minimise the transfer of vibration to other surfaces.
2. Resonance of the various parts of the machine and plates should be stopped. This can be achieved by damping material or increasing their stiffness or weight.
3. The noise may be reduced by reducing the area of the vibrating parts.
4. The noise-radiating characteristics of the noise source can be changed by changing the fixing positions of the vibrating surfaces.
5. Noise from air movement should be reduced using acoustically-absorbent linings in ducts or by using an 'acoustic tunnel' to absorb the noise from the unit.

It is less common in modern control rooms to find noisy items such as impact printers. Where these or other noise sources are found, however, a common method of noise reduction is to prevent the spread of noise from the source by using an acoustic enclosure. This is often also a sufficiently good way of preventing noise disturbance. Any noise-reduction enclosure installed should not interfere with the control room work to be carried out. Most types of enclosure, for example, hinder visual tasks. Enclosure of printers used in computer installations often makes it considerably more difficult to read the printed paper. Such a situation may sometimes be ameliorated by installing special lighting inside the enclosure. Enclosures should be provided with sufficiently large windows that have some form of antireflection treatment.

The walls between ordinary offices have a sound reduction effect of about 30 dB. The walls of control rooms should be provided with at least 35 to 40 dB reduction, and restrooms where privacy is required should have a noise attenuation of around 50 dB.

One general background noise that often occurs in control rooms is the noise from fans used for ventilation of both the computer equipment and the room. In the worst cases, the noise from the fans can rise to over 70 dBA. The equipment and the control room itself should be ventilated to ensure that noise levels do not exceed 35 dBA. This is usually easily arranged by providing the ventilation ducting with internal sound-insulating material or in some cases by installing special acoustic baffles.

Infrasound—that is, sounds at frequencies below the audible range—may cause subjective problems and perhaps also have a negative effect on the level of alertness. Infrasound may occur, for example, in air-conditioning systems, water turbines (such as in a power station), silo installations, and dryers.

7.4.5 VIBRATION

In control rooms, hand-arm vibration is rarely a problem. But significant levels of whole-body vibration (WBV) can often occur where the process vibrates and the control room is attached to the same structure as the process. This can occur, for

FIGURE 7.10 Control room mounted on a primary crusher in a quarry. The WBV-weighted vibration on the floor of this control room exceeded 1.15 ms^{-2} on one axis. (Photograph by A. Nicholl.)

example, where shakers are used in the process, or where a control room is mounted on a large vibrating machine such as a mobile crusher in a quarry (Figure 7.10). Vibrations from the process are transmitted to the floor of the control room, and hence to the operator's chair.

As with noise, regulations produced under the EU Physical Agents Vibration Directive (2002) limit the average daily WBV exposure. The limits are based on comfort criteria. Vibration in control rooms is usually of low frequency; in the case of the quarry crusher, there were also random jolts as the rock was crushed.

The best solution to the problem of vibration is to isolate the control room from the process, a solution best introduced at the design stage of a project. Once the control room is installed, seats specially designed to reduce the seated operator's vibration level may be needed in order to reduce the transmission of vibration to the operator. Such seats need to be designed for the range of frequencies found in the vibration, and also to be adjustable for the weight of the operator. They also need to damp vibration in the lateral and fore-and-aft directions as well as the vertical direction.

Whole-body vibration can cause a range of musculoskeletal problems. It causes the muscles to tense up in order to counteract the shaking of the body, leading to fatigue and stress, and it can aggravate back pain. This can lead to a reduced level of attention on the task. WBV can also make instruments, labels, and documents difficult to read, where either the operator or the instrument, or both, are vibrating.

REFERENCES AND FURTHER READING

Atherley, G.R.C., Gibbons, S.L., and Powell, J.A. (1970). Moderate Acoustic Stimuli: The Inter-relation of Subjective Importance and Certain Physiological Changes. *Ergonomics* 13, 536–45.

Beranek, L.L. (1954). *Acoustics.* New York: McGraw-Hill.

Boyce, Peter R. (2003). *Human Factors in Lighting* (2nd ed.). Boca Raton, FL: CRC Press.

Broadbent, D.E. (1957). Effect of Noises at High and Low Frequency on Behaviour. *Ergonomics* 1, 21–29.

Broadbent, D.E. (1963a). Some Effects of Noise on Visual Performance. *MRC Reports in Psychology*, no. 15. Cambridge: Medical Research Council.

Broadbent, D.E. (1963b). Differences and Interactions between Stresses. *Quarterly Journal of Experimental Psychology* 15, 205–11.

Chapanis, A. (1969). *Research Techniques in Human Engineering.* Baltimore: Johns Hopkins Press.

Cherry, E.C. (1959). *On Human Communication.* Cambridge, MA: M.I.T. Press.

Davis, S.W. (1955). Auditory and Visual Flicker-Fusion as Measures of Fatigue. *American Journal of Psychology* 68, 654–57.

EU. (2002). Directive 2002/44/EC of the European Parliament and of the Council of 25 June 2002, on the Minimum Health and Safety Requirements Regarding the Exposure of Workers to the Risks Arising from Physical Agents (Vibration). *Official Journal of the European Commission,* 6 July 2002.

Fanger, P.O. (1970). *Thermal Comfort.* Copenhagen: Danish Technical Press.

Grandjean, E., and Etienne, P. (1961). Pupil and Time Effect on the Flicker-Fusion Frequency. *Ergonomics* 4, 17–28.

Griffin, M.J. (1990). *Handbook of Human Vibration.* Oxford, U.K.: Elsevier.

Hopkinson, R.G., and Collins, J.B. (1970). *The Ergonomics of Lighting.* London: Macdonalds Technical and Scientific Publishers.

HSE. (2005). *Controlling Noise at Work*. The Control of Noise at Work Regulations, 2005: Guidance on Regulations L108, Health and Safety Executive. London: HSE Books.

Mansfield, Neil J. (2004). *Human Response to Vibration*. Boca Raton, FL: CRC Press.

Weston, H.C. (1962). *Sight, Light and Work*. London: H. K. Lewis.

Van Cott, H., and Kinkade, R.G. (1972). *Human Engineering Guide to Equipment Design*. Washington, DC: American Institute for Research.

Part IV

Case Studies and Applications

8 Industrial Applications and Case Studies

Erik Dahlquist, Brian Hunt, and Toni Ivergård

CONTENTS

8.1 CASES OF CONTROL CENTRES, DISPLAYS, FUNCTIONALITY, AND LAYOUT

A hundred years ago, control rooms looked beautiful with all copper and brass shafts and wheels, and mechanical displays. During the 1970s and 1980s computerisation and standardisation became the norm. Every part of an industrial plant had its own control room. Everything possible was displayed and logged, and enormous amounts of data were stored in files. Proportional, integrational, differential (PID) controls were implemented on a large scale, and alarms were installed on an equally large scale (Figure 8.1).

The final years of the 1990s and the beginning of the twenty-first century saw a centralisation of the control room operations so that the number of necessary control rooms was often reduced to only one or two in a complete industrial plant. No

203

FIGURE 8.1　The control room at Västerås old power plant (around 1920).

longer are operators handling a small part of the plant that they know in great detail. Instead they work in teams who cover the complete plant, or at least major parts of the plant.

This change in control room focus also brought about changes in the need for much more automatic controls and decision support tools. It is not enough to collect a lot of data. What is important is to be able to make use of these data to enhance the production, minimise the quality variations, minimise shut downs, avoid environmental harmful emissions from the plant, reduce risks for accidents, and so forth. Commercial pressures mean that all this should be done with fewer people. From a situation where the focus was on local displays for a smaller part of the plant, organisations now need a control room where small displays interact with large displays and with possibilities for conferences to share and exchange information, and a pleasant environment for the operators.

One of the leaders in this area is the multinational conglomerate Asea Brown Bouverie (ABB). They have created a concept where a few separate computer displays on a desk can be lowered or raised as needed by the operators. Above the desk are a number of large displays. There is no possibility for the operators to walk between the desk and the large screens, as this might block the visibility of the screen for the other operators. ABB realises that it is also important to have information displayed on such a scale to be readable from a distance. When a particular operator is working with data in the smaller displays and needs a second opinion from colleagues, the operator can bring up that part of the process on the large display. Everyone in the room with relevant knowledge can contribute to the discussion. When the operators need more information to solve a problem, the technology can also display images from cameras or videos situated within the plant.

On the smaller displays it may be possible to display, relatively easily, a whole section of the plant from one end to the other. This is much better than having a series of very small pictures to provide an overview. Data can also be combined by displaying the process flow sheet on the large screen while trend displays are displayed on the small screens (Figure 8.2).

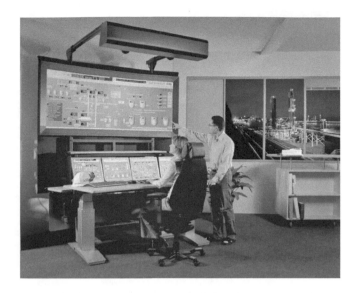

FIGURE 8.2   Modern control: The demonstration system at ABB.

When the operators need to sit and discuss problems, they are able to do this alongside the displays at conference tables (Figure 8.3). These tables intended for conference meetings have been designed into the layout of the control room and are thus adjacent to the data displays. Restrooms and similar facilities are close so that operators can be brought in quickly when needed. An example of such a modern control room design is shown in Figure 8.4.

Formerly, everyone using a computer screen received the same information from the computer system (Figures 8.5 and 8.6). Nowadays, a new trend is to personalise the displays. This is important because each individual person or team should receive the information relevant to his or her particular role in the plant operations. The maintenance people should get the information relevant for service and maintenance, the production manager should get the overall picture that is relevant for the planning of the whole process, and the process engineers should have the possibility to go into deeper depth of the process information system to develop the process or diagnose the process performance. For the management it may be no technical information at all but only some key information about process stability, mainly focusing on production level, quality, material flow, and similar tasks, as well as order booking and deliveries.

This differentiation has both advantages and disadvantages depending on the perspective. It may give the possibility to differentiate the information so that different operators have their own displays. This has the advantage to make each operator happy, but the disadvantage is that it may be more difficult to operate the plant in the same way for different shifts if everyone is allowed to modify the information he or she should retrieve. So decisions must be made on a strategic level about what differentiations may be allowed and what needs to be standardised.

FIGURE 8.3　Large screen used to show details for a large group so that discussions can be made to solve problems together.

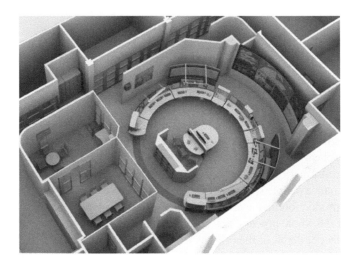

FIGURE 8.4　Modern control room layout at ABB.

8.1.1　Some Experiences from the Field

Starting in 2004, a major rebuild was carried out at a pulp and paper mill, Korsnäs, in Gävle, Sweden. The company closed thirteen control rooms and replaced these with only two new control rooms. At the same time, the staff was reorganised so that the complete production process became the responsibility of only one production team. Similarly, only one manager was responsible for the complete production

FIGURE 8.5 Siemens system at Billerud, Skärblacka: Large and small screens combined.

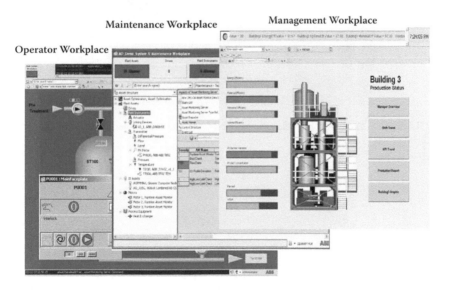

FIGURE 8.6 Personalised workplace with filtered information.

process. Functional departments were replaced. So instead of having departments with control engineering, maintenance department, and so forth, these functions were integrated in the new production team for the fibre line, and respectively the paper mill and the recovery area. The production teams directed how the operations should be performed by all shifts. This avoided the problem of different shift teams operating in different ways. In this new way of working, the goal is to identify the best possible way and then every team should follow that direction.

In 2005, Billerud, an integrated pulp and paper mill located in the town of Skär-blacka in southwest Sweden, rebuilt its power plant. Using a Distributed Control

System (DCS) from Siemens, Billerud reconfigured its control rooms in a similar manner to those at ABB. The emphasis is on clear, unambiguous process data. Now, data can be displayed on large display screens to complement the data displayed on smaller monitors. As necessary, the operators can view data from different perspectives and with different emphasis. Operators can also combine their specialised knowledge in collaborative working and decision making. At Billerud, the process displays are configured in a slightly different layout, but otherwise the structure is similar. Other, but alike examples are shown in Figure 8.7 and Figure 8.8.

FIGURE 8.7   Old control panels are removed from the Mälarenergi control room.

FIGURE 8.8   After the rebuild in 2005, the new control room has a combination of large displays on the wall and a number of small screens for the operators.

As shown in the figures, the operators have an overview picture of the plant on the large screen and to the right there is the alarm list displayed on such a scale that it is easy for the operators to read. The small screens display details of the process. Operators choose between the two types of displays as appropriate. As in the case with the ABB system, there is no possibility for an operator to walk in between the small screens and the large screens and restrict the view of colleagues.

Mälarenergi is a power company with its central power production plant in Västerås, Sweden. In 2005 Mälarenergi made a major revision of the plant control room. The DCS system is primarily an ABB Advant system, although older versions also exist. Until 2005, Mälarenergi also used the old 'pulpit' type of controls with hard-wired controls and displays in parallel to the more modern computerised controls. As part of its plant upgrade, Mälarenergi modernised its control room to give a common working environment for the operators and process engineers. At the same time, the hardware types of displays were removed, as shown in Figure 8.7. Instead the designers installed a combination of large screens and small displays. This new configuration gave the operators a much more modern environment according to guidelines recommended by ABB. Here each operator station has four screens kept together, as seen in Figure 8.9.

After the project manager at Mälarenergi and his team had gained experience with the new system, he summarised its advantages and disadvantages. The advantages of the new control room and the new computer system are:

1. The hardware has much higher capacity.
2. All parts of the process and all systems have the same layout, which makes it easier to recognise what is happening.
3. You can work only in a Windows environment and do not need to keep up with UNIX, OPEN-VMS, or other platforms any more.

FIGURE 8.9   Each operator workstation has four screens to see both overview and detail.

4. The new aspect system introduced by ABB makes it easy to find different types of information for specific equipment or parts of the process without having to navigate from far away.
5. You have direct access to the Internet and other online resources.
6. You have a major possibility to customise your own displays with the information you really need.

Disadvantages with the new system compared to the old pulpit system are:

1. Every time there is a new version of software, for example, there is a need for upgrading. This was not needed before the controls were computerised.
2. The software solutions are probably more sensitive compared to the old hardware solutions, although the software solutions also have many new features.
3. You need to make lots of movements of the mouse to navigate the display system.
4. Some of the older operators find it more difficult to learn the new system with more functions
5. The overview of the process is not as good as with the pulpit system.

Even if there are some drawbacks with the new computer-based systems, it is also obvious that the functionality is much higher. New types of functions for advanced diagnostics and control can be implemented. Also, with the large screens on the wall, the overview is significantly better than when there were only small screens on desks.

8.1.2 FUTURE TRENDS

What we can see as a future trend is more of the 'pump up' types of functions. As long as a process is going well, there is no need for information about it. But when something starts to go wrong, an early warning should show up with all information relevant for diagnosing the problem. There may also be a need for more advanced types of graphics. If you are out in the plant you see everything as three-dimensional (3-D), but on the display everything is two-dimensional (2-D). To make it easier to see where something is situated, such as sensors, it may be good to have 3-D displays—for example, an image that zooms in on a sensor when a fault is detected. Another future trend will be the display of new types of data and information than before. To date, the majority of information has only been available on a 'standalone' basis. Trends of change can be observed from separate images. These can be filtered to reveal the core of important data. This allows particular segments of data to be compared.

In the future it will be possible to produce different types of key performance indicators (KPIs) for different parts of the process. It will also be possible to follow these step-by-step to understand the process and the symptoms of any faults that build up over time. Operators will be able to construct different types of process models and then to fine-tune these online using new process data, after verification of the quality of this collected data.

8.1.3   THE CONTROL ROOM AS PART OF AN
INTEGRATED PROCESS DESIGN: STORA ENSO

Stora Enso is an integrated paper, packaging, and forest products company based in the west of Sweden.* The group has 46,000 employees in 40 countries. In 2005, Stora Enso made total sales of €13.2 billion. Founded at the end of the nineteenth century, the company now exports 85% of its production. With a production capacity of 1 million tonnes annually, the company's paper mill is one of the largest in the world. Stora Enso's paper mill has a thoroughly developed control room that is impressive for its merging of environmental design and architecture. Stora Enso regards fundamental design as important. The company has thus expended much effort on its organisational design with a matching architectural design.

The control room is an integrated feature of the papermaking process and has been designed in such a way that, although physically separate from the manufacturing area, it seems to be a part of this facility. This is achieved by aligning the ceiling height of the control room with the ceiling height of the paper mill hall. The high ceiling height and the ceiling height compatibility encourage the operators not only to feel physically part of the paper mill but also psychologically so. This design makes the environment amenable to operator control of the manufacturing process, while allowing verbal communication between the operators in their work. Sensitive design has reduced to a minimum the ambient noise and vibration. This has been achieved by the use of double glazing, and special locks and door-opening mechanisms. Excessive noise and air pollutants are thus prevented from entering the control room. A specially constructed floor neutralises vibrations from the adjacent manufacturing area and prevents vibrations from disturbing the control room work and its electronic equipment. Similarly, lighting design is rather sophisticated. All lighting is glare-free. The positioning and direction of lighting fixtures add to the psychological feeling of that operators and their work environment are a part of the paper mill hall. Furniture design, including chairs and desks, and the positioning of displays all meet reasonably good ergonomics standards.

A key feature of Stora Enso's organisation design is the support of control room operators who are 'all-rounders' and skilled in multitasks. Specialisation is avoided as much as possible, as this restricts an operator's skills and competencies. Control room operators are obliged to work both in the control room and out in the paper mill. Their work tasks encompass process servicing and some aspects of maintenance and repair work. Control room operators are well-educated engineers; currently, they are discussing the need to employ some specialists in information and communication technologies (ICT). In this high-tech environment, supervision, maintenance, and development of information technology (IT) are of increasing importance.

Control room operators are also involved in development work. To a large extent this type of work is done in meeting rooms that are an integral part of the control room. Stora Enso also provides continuous training in different job-related areas. Although continuous learning is an everyday feature of the operators' work, there are no special simulation facilities in the existing control system. The computers that

---

* Data for the case study were generously provided by Stora Enso in face-to-face interviews with selected managers, from company documents, and from the company's Web site.

FIGURE 8.10   Overview of the 'high ceiling' control room at Stora Enso.

are part of the process control system are not directly used in any systematic way for training and development.

An area for future improvement is the lack of availability of large-scale overview displays. Today the operators use static, paper-based drawings of the whole process (observable in the right-hand background in Figure 8.10). Stora Enso has investigated this possibility but say they are unable to identify any suitable products currently available. As soon as the market can offer some good solutions, they plan to develop large-scale overview displays for their own purposes.

8.2   CONTROL ROOMS FOR POWER GENERATION AND DISTRIBUTION

Hydroelectric power stations produce the vast majority of Sweden's production of electricity.* The system for production consists of reservoirs and dams in the upper parts of Sweden's network of very large rivers. Reducing the upper water levels in the dams produces the static energy used to generate dynamic energy. Turbines convert this dynamic energy into electricity via electrical generators. After being transformed into an appropriate voltage, the electrical energy is distributed via a vast network of power grids throughout the country. In this way, the electricity is carried from the remotely situated dams to populated areas and industries. Until fairly recently, the cabling was mainly mounted on very high-level pillars. Nowadays, the grids nearby populated areas are channelled mainly underground. High volumes of electricity are distributed via underwater cables along the coastlines or across the seas.

Power generation—including the dams, dam gates, turbines, and electrical generators—is supervised and controlled from special control centres, including control rooms manned by people. In the past a large proportion of the control and logistical

---

* Data for the case study were generously provided by Vattenfall in face-to-face interviews with selected managers, from company documents, and from the company's Web site.

manoeuvring was done manually. Successive technical devices have taken over most control functions. Mainly responsible for routine monitoring of the system, the operators have a key role in intervening in order to avoid malfunctions and accidents. Over the past decades there has been a separation of the control rooms for function of electricity production and the control of electricity distribution.

A central control room monitors and controls the overall production and distribution of electricity, including the import and export of electricity for the whole country. This centralised national control centre could handle (at least in theory) all of the monitoring and control functions for the whole country. In practice, it is thought wise to maintain a number of regional control centres to supplement the local centres and the control rooms for the production of electricity. In 2007, there were only three regional control centres. In the future these might be reduced to one or two.

In the past the people working in control rooms and at the local control centres placed a strong emphasis on the practical value of having personnel close to key parts of the distribution grid, including the power production control rooms. Local knowledge, in the form of tacit knowledge, is of great value to reduce disturbances when there are major fluctuations in electrical supply, such as during large storms. When parts of the distribution grid are disrupted and become incapable of electrical distribution, there is a risk that the problem will create a local backlog. This could result in an overloading of the remaining parts of the network, and might disrupt even larger areas of the network. It is not difficult to imagine the knock-on effect from a minor problem spreading into larger regions, causing large numbers of people and organisations to be without electricity.

Thus one important area of technical and human-technology risk management is to define the optimal combination of the automated system in relation to human operator support. In the past, a weakness of the system has been in fully understanding the actual capability of the human operators. In this context, a classic dilemma is local knowledge versus centralised decision making.

The enormously large dams represent another area of technical risk management. In fact, all of the different types of related technologies needed to generate and distribute electricity involve degrees of technical risk. The automation system as such also represents a degree of risk. The related computer system and its software is, to an increasing degree, an important risk factor. To cover this part of the system, an appropriate risk management capability is needed. This capability also includes available skills and knowledge among the control room operators.

Other sectors of the 'control room industry', such as the paper and pulp industry, are considering the use of operators specialising in software engineering on twenty-four-hour duty rosters.

In the past control rooms were more or less allocated near to most power generation sites. Nowadays, it is only necessary to have a few control rooms cover a large number of power generation units and related dams. One consequence of this development is the obvious reduction in the access to local knowledge among the operators. One control room might control an area of many hundreds of kilometres. An additional problem is that supplementary human visual contact is needed to ensure the more direct supervision of dam gates and other areas. In this new context, supervision needs to rely on supplementary use of locally placed video cameras connected to related displays in the control rooms. Today, the quality of video display techniques is of rather high quality, but there remains the issue that the video images are two-dimensional whereas the real world is at least three-dimensional. The fourth dimension (time) has also to be taken into account. For example, an operator might have difficulty perceiving slow changes in the environment such as a slowly increasing water level. Even in real life this can be difficult. The on-screen technical representation of the real-life environment might make the changing situation even more difficult to decode. This does not matter too much if the operators are well acquainted with the form and function of the real-world situation and have good local (that is, individual) knowledge. Thus equipped, the operators can to some extent mentally visualise the third dimension and add this picture to what they see on the screen. To some extent, they can also partly (but not fully) visualise the fourth dimension, time.

An additional problem is that there is also a need to add on different forms of security displays to control illegal entry via gates, doors, and other access points. A big difference today compared to previously is that there are increased numbers of video displays to handle the control and supervision of all stages in the power generation process. One reason for these changes is the increase in the numbers of threats by groups with a politicised agenda. However, when localised (individual) knowledge is missing, the task becomes even more difficult.

Some decades ago, research into control room work and automation involved intensive debate about the need for an overview of the function of the different subsystems in relation to the overall system. The old types of control panels frequently consisted of functional drawings of the related process, including analogue (frequently online) instruments for important variables and also some controls (knobs and dials for making adjustments and on-off switches). The positioning of the controls, visible to the operators, meant that the controls as such also became displays of the current control status. In the era of intensive computerisation of the control process, the old fixed panels were removed. The old inflexible hardware was replaced by keyboards and cathode ray tube (CRT) displays. Obviously the new solutions were much more flexible and thus it was much easier to update the changes in the process.

However, at the same time, the advantages of having a good overview of the system were lost. A major problem is the fact that the display modality of control positioning has disappeared. This means that the operator receives no immediate visual feedback from the positioning of the controls' (knobs and levers) current status. A striking example is the use of a manual gear stick of a truck. By looking down at the position of the gear stick the driver receives immediate visual feedback. For a practiced driver it might be sufficient simply to feel the position of the gear stick. This kind of visual feedback is lacking for the operator who is only using a keyboard. This is the negative functional trade-off from the flexibility offered by the keyboard. In the future this deficiency has to be resolved. It might be difficult to return to the old types of controls in the forms of knobs, dials, levers, and so forth. But it might be useful to include into the display the control positionings or control status. In some special cases it might even be prudent to use 'old-fashioned' types of controls.

In many modern control rooms (not only in power control rooms), the design of the information presented to the operator seems to distance itself from the real-world process that it should represent. It is important that the information presented gives the operator the correct and optimal feeling for the process and its current status. This should be the main role of an overview process display. This display should continuously support and update the operator's mental and tacit model of the ongoing real-life process.

It is a risk to put too high a degree of emphasis on local desktop displays. The desktop should give detailed information. Frequently, an operator works with several desktop displays in parallel. It is important that the different displays represent different modalities. The overview display should show a representation of the status of the main process.

Overall, existing displays seem to be orientated more towards rather static information in the form of different types of important indicators—for example, numerical. Ideally, such information should be process and dynamic orientated in a graphic image form. New display techniques for macro presentations of overall functions of processes could probably improve the operators' understanding and ability to intervene rapidly and appropriately. Investments in these types of new displays are most likely very cost effective.

Earlier in this handbook, Figure 6.14 showed an example of a layout for a control room. A workplace is placed centrally with access to a display unit, keyboard, and special controls. There is also space on the central workspace for writing and for various reference books. The process, which is assumed to be large, is displayed on the overview panel, mainly in the form of static information. Inserted into the display is certain dynamic information, such as the settings of the more important controls, critical readings, or deviations from the set values. The overview board is designed so that all parts of it are at more or less the same distance from the operator. There is also a more traditional office workplace in the control room, where various special tasks can be carried out during daytime.

In a control room where there is an overview panel and several people are working, it is important that this is positioned so that everyone can see it.

A later section (Chapter 11) of this handbook describes the design of learning and creativity at work. In most control rooms for power generation the operators are underloaded. In such a situation, the work tasks do not prepare the operators for the more demanding tasks that occur, albeit infrequently. There might be hours or days between major highly demanding tasks. The main reason for having operators in a control room is to handle these high-peak critical tasks. It is important for the operators to prepare themselves for these essential situations. One way is for the operators to use the existing computer systems to simulate possible breakdowns or other types of critical incidents. This will allow the operators to improve their ability in handling disturbances. They will also obtain improved knowledge and insight into the function of the control system and the processes controlled by the automation. As a part of this simulation training, the operators themselves can contribute to the development of the system and its automation. If the operators become proactive in the development of their own workplace and its related tasks, they will also become much more motivated and creative.

8.3   TRADING OF ENERGY AND ENVIRONMENTAL ISSUES

In this case study we describe the control rooms and systems employed in the trading and brokerage of energy.* We describe a brief history of the growth of trading and discuss the roles of trading in current-day commerce. We use a contemporary case study to exemplify the trading of energy and associated environmental issues. Using this background we discuss how control room design and management contribute to an amenable workplace environment in which traders can work. In our conclusions we offer a number of proposals for control room design and management.

Our case study example is Vattenfall, a Swedish energy generating and trading company. Vattenfall (the name means waterfall) traces its historical roots back to 1899 when the Swedish Parliament passed legislation to use water rights for industrial purposes and set up an organisation called the Waterfall Committee to manage this natural resource. Nowadays, energy trading is a key feature of Vattenfall's business model and is a fundamental part of the company's value chain. For several decades, Vattenfall's energy production and physical flows have been controlled and monitored in the company's national, regional, and local control rooms. Separately, other types of the company's control rooms monitor financial and informational data flows. Together these control rooms monitor the diverse information needed for the business processes and help optimise energy supply and energy demand. However, the scope of financial trading is much wider than, and independent of, the production and distribution of energy. This form of commodity trading exemplifies a business environment that depends heavily on highly competent financial traders processing diverse information in rapidly changing real-time scenarios. The traders rely on accurate presentation of salient data in order to understand evolving market movements.

8.3.1   GROWTH AND DEVELOPMENT OF THE
TRADING AND BROKERAGE INDUSTRIES

Traditionally, brokerage and trading services are associated with the financial markets. Early systems of credit and debit were practiced in Cairo from the eleventh century as the city benefited from being at the crossroads of the spice route from Asia to Europe. Venice, which had been an international trading centre for several centuries, rose to prominence in the twelfth century. Capitalising on its position as one of four city-states in Italy and its strategic location on the Adriatic Sea, Venice was ideally placed for trading with Byzantium and Asia Minor, which were at the western edges of the Silk Road. In 1295, the Venetian trader and explorer Marco Polo (1254–1324) returned from his travels with unbelievable tales of distant lands and opportunities for trade with central Asia and beyond.

In northern Europe from the twelfth century, debt trading and systems of brokerage were features of commerce in France and Flanders (now parts of Belgium and Holland). In the wake of burgeoning mercantilism, bourses opened in Bruges (1309) to service the trade in wool and cloth and in Antwerp (1460) to finance the trade

---

* Data for the case study were generously provided by Vattenfall in face-to-face interviews with selected managers, from company documents, and from the company's Web site.

in diamonds. The following century saw bourses opened in Lyon (1506), Toulouse (1549), Hamburg (1558), and London (1571). These were followed, in the seventeenth century, by bourses in Ghent, the centre for the trade in wool and cloth weaving, and Amsterdam, a centre for maritime trade. The growth of overseas trade links was accompanied by markets in commodities and futures. In 1773, London traders who had been meeting informally in the coffee houses that sprang up were able to meet in a building designated the Stock Exchange.

Trading in commodities such as natural resources has an illustrious heritage. Around 2500–2000 B.C.E. Sumerian traders in Mesopotamia (now Iraq) carved on tablets of clay records of their commerce in livestock and precious natural objects such as stones and seashells. Called *cuneiform*, this is said to be the earliest system of writing.

8.3.2 TRADITIONAL MODELS OF TRADING AND BROKERAGE AND EMERGING TRENDS

The introduction of the free-market economy from the end of the nineteenth century and the past hundred years has had a dramatic impact on trading and brokerage. Trading is conducted to generate profits for individuals or a single business entity. Brokerage is conducted from an intermediary position between buyers and sellers. Profits are generated by differentials in prices between the buying price and the selling price. In the late twentieth century, brokerage and trading became more closely integrated. This was partly driven by developments in technology and changing global geopolitics. With the breaking down of national borders in the interests of trade and the increased power of electronic media, the 'borderless world' became a reality. In an international and global context, complexity gave way to relative simplicity. Databases continue to expand in size and scope. Contextual complexity is reduced to standardised commoditisation. The total flows of information and its ready accessibility sets an open environment for globalisation.

In the European Union (EU) arena of energy and related environmental trading, the main objective is to optimise the access to environmentally friendly energy at the lowest possible cost. This process of trading is obviously also dependent on the availability of a physical infrastructure. This structure of logistics includes all areas of distribution of commodities but the infrastructure for production and distribution of electricity is particularly important.

The workplace environments of the trading and brokerage personnel increasingly resemble the control rooms found in industrial settings. At the same time, trading and brokerage activity in different contexts have become more similar. At times it is difficult to identify the optimum environment and setting for any given control situation. Free-market concepts are governing the principles and methods being used in this daily work. Increasingly, different areas of trading and brokerage use the same types of databases. The concepts of corporate social responsibility (CSR) and other forms of ethics have been developed in these types of industries. Brokerage and trading can be seen as industries of their own accord. Today financial trading and brokerage can be found in many different industry sections and with many different applications. In the financial services industry, one area that is growing very rapidly is related to trading in energy and related environmental issues.* In Europe, financial trading and services in the energy sector have been harmonised into one conjoined market for all EU countries and many EU-associated countries (such as Norway). Obviously, in this financial sector, different forms of ethical rules and regulations, such as those regarding insider trading, become very important.† Many different actors have been established across this new market. As an example of this, we have been looking closely at the Swedish company Vattenfall.

8.3.3 TRADING AND BROKERAGE OF ENERGY: VATTENFALL

In 1909, the Royal Waterfall Board (Vattenfall) was founded to generate electric power from the natural resources of Sweden's waterfalls.‡ The company grew rapidly, building coal-fired plants for generating hydropower and developing a unified grid system to connect different parts of the country. By the early 1950s the whole of Sweden was interconnected in a power generation network. In the 1960s, the network spread to neighbouring Scandinavian countries. In the light of increased concerns about overusage of water resources, Vattenfall developed heavy water reactors and subsequently developed light water technology using enriched uranium. Sweden's first nuclear power reactor became operational in 1972. By the mid-1970s, Vattenfall owned seven of the twelve nuclear reactors in Sweden. However, by the end of that decade political realities saw the scaling back of nuclear energy programmes. In 1992, following the deregulation of the Swedish energy industry, Vattenfall became a public limited company.

Encouraged by EU directives that opened up national energy markets to cross-border competition, Vattenfall successfully pursued commercial opportunities outside Swedish borders. To thrive in this new spirit of competition Vattenfall needed to change its business model and corporate mind-set from one of cooperation to one in which the company was proactively competitive. Vattenfall thus acquired the Finnish network company Häsmeen Sähkö and opened an office in the German port of Hamburg. International expansion quickly followed. The number of employees in the company's operations outside Sweden increased from 10% to 40%. With the creation in 2002 of Vattenfall Europe, 77% of the company's employees now work outside Sweden. Benefiting from its historical roots in Sweden, Vattenfall traditionally developed its main markets in Scandinavia. Today the company's market presence in Germany is of equal size and the company is developing its market presence in Poland. Vattenfall (as of December 2006) employs over 32,000 employees, of which 493 people work in energy trading, finance-related activities, and other services shared by the group overall.§ Sales for the year 2006 amounted to €16.1 billion. Sales within Nordic countries were €5.3 billion, sales within Poland were €0.99 billion, and sales within Germany were €7.7 billion. Electricity produced by the

---

* See, for example, James Kanter, New Climate for Emissions Trading Turns Greed into Green, *International Herald Tribune*, 22 June 2007.
† See, for example, James Kanter, Stiff Rules Proposed on Carbon Trades, *International Herald Tribune*, 30 June–1 July 2007, page 14.
‡ Data from http://www.vattenfall.com, and from personal communication with managers at Vattenfall.
§ 'Personnel employed' is calculated on the basis of man-years.

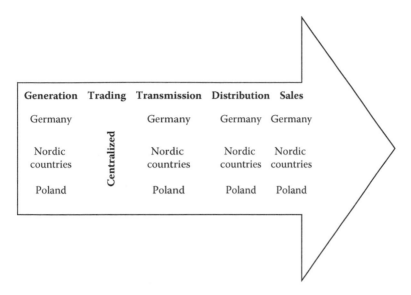

FIGURE 8.11   Role of energy trading in the Vattenfall value chain.

Vattenfall group is sold to 4.9 million customers: 0.9 million in the Nordic countries, 1.1 million in Poland, and 2.9 million in Germany.

8.3.4 VATTENFALL AND THE ROLE OF TRADING

Figure 8.11 shows Vattenfall's business concept and the role of trading in the Vattenfall value chain. As can be seen, the value chain covers generation of energy, trading, transmission, distribution, and sales. From the perspective of its place in the Vattenfall organisation, trading is a key centralised function, a fulcrum between supply and demand. However, the functionality of trading processes is virtual and reaches markets far away from the cabling and pipes carrying the traded commodity. The total traded volumes are much larger than the volume of energy generation, transmission, and distribution within the value chain.

Vattenfall's vision for financial trading is to be the risk management centre for the organisation. In this context, risk refers to financial risk. The mission of the financial trading function is to translate trading competence into additional value for Vattenfall's customers. Vattenfall's customers are the boards supplying power-generated services such as electricity.

By far the largest traded commodity by volume is electricity. Trading is not limited to hydroelectrical commodities, hard coal, gas, oil, and other traditional fuel sources. An important part of the trading is related to emissions into the environment (measured as CO_2 greenhouse equivalents). Different forms of national and international regulations such as the Kyoto Protocols and the equivalent EU regulations need also to be considered. Other factors to be considered are international transfer pricing and exchange rate fluctuations, as errors in these areas can be costly.

8.3.5 CONTROL CENTRES FOR ENERGY AND ENVIRONMENTAL TRADING

Mankind needs to find ways to replicate the processes in which nature by itself handles production and disposal of waste materials into a positive circle and balance. Individuals and local communities will have important roles in creating this form of balance and harmony with nature. They also need to be supported by government laws and regulations and possibly also environmental 'courts'. In the global arena the implementation of environmental trading as a part of energy trading might have crucial and positive implications.

Trading in this area is, in other words, a very good example of a very complex information-processing environment. It includes most diverse areas of traditional trading from different sectors. It also uses databases from many different sources such as meteorological services, environmental information, and foreign exchange data, as well as physical resources such as transportation (for example, oil tankers, liquefied petroleum gas [LPG] tankers, and so on). It also needs close monitoring of ongoing and evolving situations in energy commodities markets. In addition, this form of trading has to be connected to available information sources of energy production and distribution. All these types of information need to be available online. Critical data and information must be presented in different forms of overview displays that simplify for the traders internalisation of their understanding of the ongoing financial and economic processes. In practice, the traders also need access to a number of smaller displays connected to different kinds of computer systems, including useful computational models. The logic behind all these forms of information presentations must be such that they support and enhance the operators/traders' understanding of the ongoing process in the market (see Figure 8.12).

Over the past twenty years, these types of models have been studied and researched in various types of industrial process control. In this type of situation we have a good understanding and knowledge about how to build up information displays and related controls to support the operators' tacit and explicit models of the ongoing processes. As the industrial specialist practicing in the control centre setting, the operator is continuously building up and improving and handling the process. In financial systems we don't have this form of experience about the critical issues related to the control centre work and its related processes on the financial and trading environments. Today, there is a reliance on the presentation of rather simple key information and data. In future scenarios, it might be expected that overview displays will be designed in relation to the economic theories related to this form of trading and also adopted to the education background of the operators and the learning facilities available. (See Chapter 11, which describes models of learning and creativity in control centre work.) The potential for development in this area is enormous, and the economic gains from this type of development could be very large indeed. An optimal design of the control centre for this type of trading could add skill and ability to the operators. And it will also add to the continuous development of the computer systems and computer software used in this context.

FIGURE 8.12 Energy and carbon trading at Vattenfall, Stockholm.

From *ad hoc* studies of a few control centres in this area of financial services, it can easily be concluded that there can be improvements in the job organisation, physical working environment, ergonomic layout of the workplaces and design of furniture, visual environment, displays, controls, and so forth. The cost for these additional improvements is marginal compared to the turnover handled by the operators/traders. Today, the operators in this type of workplace are fairly young and most are reasonably physically fit. They are normally not expected to suffer or complain about minor deficiencies in their physical or organisational environment. However, it only needs one of the top traders to get problems in the lower back or (even worse) the neck, which would make him or her more or less incapable of making rational decisions due to severe pain. This type of situation can very easily be prevented. We have decades of experience of ergonomics design that creates a smart-looking and healthy work environment. In a small way this type of environment will also add positively to the daily working performance and work outputs. A good ergonomic environment will be taken for granted and remain unnoticed. Conversely, its absence can be a 'real pain in the neck' for the trader.

More critical from the perspective of the traders might be attributes of their job design and organisational issues. These types of jobs often require more than eight hours of daily work time, and lengthy periods of sedentary inactivity can be commonplace. Periods of twenty-four-hour working may not be unusual. Traders can be expected to be rather demanding on job-related rewards and remuneration. In the long-term, highly competitive salaries in financial terms might be seen as inadequate. The characteristics of the job and organisation and the type and form of management are also of great concern. For energy and environmentally orientated traders, possible beneficial factors (inherent job-related motivators) might be the fact that they are potentially making an improvement to the global environmental situation and making a personal contribution towards reducing global warming.

8.3.6 Energy Trading, Ethics, and the Environment

Environmental issues are very deeply integrated into all types of the energy industry. This very much contributed to the fact that the industry was at the forefront of raising public awareness of ethics in their business. Successively, this led to a number of positive initiatives, and more recently there has been a fast-growing interest in areas such as environmental protection, carbon trading, alternative energy sourcing, and corporate cooperation with local communities. Increased world-wide awareness of the need for good ethical behaviour and corporate social responsibility (CSR) was even generated through negative examples such as the high-profile corporate scandals in the early 2000s. The crises that overwhelmed Enron, WorldCom, and Tyco helped changed the world of business. The interpretation of the focus of CSR varies greatly between companies and between different parts of the world. For some industrialists CSR is similar to a degree of social responsibility that is of benefit to the shareholders. Others claim that the CSR concepts encompass wider dimensions and have a value in themselves. Large and small companies should have a sincere concern for the society in which they operate.

CSR can be traced back to the foundation of the International Labour Organisation (ILO) in 1919. From that time, and continuing today, the ILO is still fighting for decent working conditions and against exploitation of labour, particularly against child labour and slave-like working contracts. These remain problem issues in most of the world's fast-growing economies. However, market economy mechanisms require a high degree of openness, transparency, and fairness. Without these fundamentals in the free market there will be important impediments to real freedoms of the market. If, for example, from within their professional life, traders and brokers possess insights of a company or a market that are concealed from other participants in the market, a competitive advantage is created based on deceit. In the energy sector there has to be a careful balance between the more technologically-orientated production and distribution flows of energy on the one side and the energy-traded issues on the other. As in other financial services, it is important to separate these two sides of the business and to ensure that they are allowed to operate independently of each other.

8.3.7 'Carbon' Trading: Sometimes It Pays to Pollute

Another important CSR issue in energy trading is the handling of environmental issues. Decades ago, environmental issues tended to be rather local (the leakage at Union Carbide in Bhopal) or at most regional (the Chernobyl nuclear disaster). Another historical example is one nation's massive use of coal fires in each room of a house—each grate burning fossil fuels in a most inefficient manner and emitting black smoke up each chimney. Fortunately, at least for the governments of Great Britain, prevailing winds were north-easterly, making the pollution a problem for the governments of the Scandinavian countries to resolve. Today, and to a large extent, environmental problems are global issues. It is, however, very important to understand that the problems cannot be resolved without thoroughly developed local measures.

In the past, environmental control was related to defining limits to the emission of all types of pollutants to the air, to water, and in the form of solid waste. The objective of the environmental control was to create a healthy environment for the lives of people and nature as such. At the first UN environmental conference in Stockholm in June 1972, there were already discussions about the right to pay for the right to pollute. The conference delegates agreed on a number of principles. These included: the protection of the environment for future generations (including the protection of wildlife and its habitat), the conservation of natural resources, the maintenance of the Earth to produce renewable resources (including the guarding of natural resources from being exhausted), and the prevention of discharge of toxic substances. At that time, there was a widespread consensus about the lack of ethics in this type of mechanism. Instead, the use of legislation, rules, and regulations was seen as the optimal method to be used in protecting the health of people and the environment.

Since then, the effect from pollution has been much more dramatic and public awareness has widened—particularly, the risk for destroying the globe's environment. This fact has put a focus on controlling the emissions of pollutions to the atmosphere. This has made it attractive to use market mechanisms as a controlling agent of air pollutants in order to restrain global warming. The free market has evolved and is now more integrated into the development of the society. In February 2006, the United Nations (UN)-Kyoto Protocols formally established the integration of environmental issues as a part of the market economy.

The new business sector of environmentally-oriented trading is the fastest growing financial trading sector. Very likely it will continue to grow and outstrip other sectors of the financial services industry. Conservatively valued in 2007 at US$30 billion, the market in managing pollution emissions is expected to grow to US$1 trillion within ten years. Carbon trading, trading in emissions of carbon dioxide (CO_2), is the current leading trading activity. Market growth is particularly fast in Europe. As the U.S. government considers fully the effects of global warming, the United States will probably soon catch up.

Presently centred on London, the market is ideally placed to respond quickly to EU pronouncements on environmental pollution. Traders are evolving systems of carbon credit exchange between companies that pollute and their greener counterparts. Governments are in a position to set capping rates for different emissions. In this context, a cap is the maximum allowed atmospheric emission from industrial activity. In Europe this would be a role for the EU Commission; in the United States, state legislatures would likely assume this role. In essence, European traders monitor the proposed pollutant caps set by the relevant EU bodies. Some industrial processes pollute more than others. Polluters can buy credits that are surplus to the needs of nonpolluters. Those who pollute, pay; those who do not pollute, earn. This is the newest and most exciting segment of the thriving financial market. In its wake is predicted to follow financing of projects that are energy-efficient and nonpolluting—a situation that is good for business and even better for the environment. The pressures of rational decisions made within the market economy can be a force for good, given the strength of the market economy globally. However, if rational decisions are made solely for the purposes of profit-seeking, the force may be more malignant.

One deficiency with the new system of CO_2 or equivalent type of trading is that sometimes it pays companies to pollute. However, this is not the intention. Using market mechanisms to guide debate and action on environmental issues brings many limitations and difficulties in application. However, market forces have proven to be a very powerful mechanism for making changes and hopefully in the right direction. In 2005, the systems of CO_2 trading built on the UN-Kyoto Protocols, whose main objective is to tackle global climate change. There has to be a combination of measures in place that are part of the macro structure—for example, integrated into a market-economy type of energy trading incorporating trading related to CO_2 equivalents concerning most different types of air pollutants and related interacting and facilitating variables. This type of macro mechanism must be supplemented by joint UN-supported declarations such as the Kyoto Protocols. But it is also necessary for nations and local communities to have their own systems to control unreasonable production emitted into the atmosphere, into the seas, and in the form of solid waste. It is important to find good mechanisms based on ecological principles in all types of man-made activities. We need to learn from nature and its ability to balance production and consumption.

8.3.8   CONCLUSIONS

There is great potential to meld emissions trading with energy trading in a symbiotic relationship that is mutually beneficial. Here the different traders would work together to achieve the maximal benefit for each market (in that same way that a fuel-saving device in a vehicle indicates the speed at which fuel is optimally used). In this newly developing but fast-growing business sector, there are desirable outcomes for ergonomic design of processes, systems, and technology. In any data-driven industry, monitoring, control, and knowledge development are critical areas. Success often is decided by design of human-technology interfaces and timely ease of access to the databases. However, the additional dimension that the activities of the traders contribute to environmental protection means a stress-laden workplace situation in which the traders routinely operate. 'Saving the planet' is also a very strong motivational factor for people. In this way environmental trading is very different from the usual concept of trading.

This does not, of course, mean that features of design can be overlooked or skimped; rather, they should be key features of this workplace. A small investment will realise large results, as we describe and discuss in our following proposals.

8.3.9   PROPOSALS

- *Design of the control centre for use by highly professional traders optimised in terms of furnishings, acoustic, and visual environments.* To avoid physical strain and, in the worst cases, health problems, a control centre for energy trading should fulfil the same requirements as other control rooms described in this handbook. In other words, the operators will have access to several local displays at their own workplace and at least one major large-scale overview display. The visual environment needs to be carefully planned to avoid glare and visual fatigue.
- *Headsets with attached microphones used to avoid overhearing or disturbing general conversations.* Obviously, background noise (e.g., from computer printers) should be reduced to a minimum. Noise absorption levels

should be very good. As in all other control room situations, in this type of control room, chair design is of critical importance. However, the chair should be more designed for active work than for easy chair-type relaxation. The latter is often used in traditional control rooms.

- *The physical design of the display and information devices, including overview displays.* This type of work uses many different types of databases. This makes it important that the trader has access to several different desktop displays. In this way, the trader can operate several databases in parallel without start-up delays. A few infrequently-used displays can be placed on a higher level, above the line of sight. It is important that an overview display presents critical information and important indicators that need to be simultaneously available for all operators. The overall display should also present a general process overview. This process overview needs to be designed in relation to the specific requirement of each trading centre. The specific requirements of displays are described in Chapter 4. The main control devices are the keyboard and the cursor mover (e.g., the mouse). However, in each trading configuration it is important to define critical control actions and critical incidents that demand control actions. This type of control needs to be accessible with minimum time delay. Preferably, this type of control should alternatively be carried around by each operator and accessible without time delay.

- *Design of software for information processing and display.* For this type of trading the traders need to have access to local computers for designing their own software or adapting existing software. This software can be used for predictions, decision making, and general statistical and probabilistic computations. Individual, personalised PCs might be the best solution.

- *Job design for traders and related personnel.* On the one hand, it is very important that the traders have as extensive a decision-making authority as possible. On the other hand, it is also important that the work is carried out with the optimal level of ethics. Some ethical guidelines must be available covering areas of insider trading and environmental issues, and computer systems should be programmed to give critical warnings.

- *Design of simulation models.* In many process control situations, it is valuable to have access to computer-based models of the controlled process. This computer model can be used to simulate and predict different states of the process. In other words it can be a useful tool for decision making. If the background model is good enough it can also be used for training and retraining, particularly on infrequent events and scenarios (in the same way as the on-board training simulators are used in the cockpits of aircraft).

- *Design of continuous learning facilities.* The simulation models described above can be one tool used for continuous training and updating of knowledge. It is, however, also important for the traders' continuous training and updating to be related to their complex area of operation (see Chapter 11).

- *Design of schemes for creative development.* In their daily work, traders are very busy and sometimes also in highly stressful situations. Highly educated and competent professionals, the traders are very competent to be the prime developers of their own development. To be able to use this potential, it is essential that they can be relieved from some of their daily workload.

REFERENCES AND FURTHER READING

Adler, Paul S., Goldoftas, Barbara, and Levine, David. (1997). Ergonomics, Employee Involvement and the Toyota Production System: A Case Study of NUMMI's 1993 Model Introduction. *Industrial and Labor Relations* 50, 3 (April), 416–37.

Carver, Liz, and Turoff, Murray. (2007). Human-Computer Interaction: The Human and Computer as a Team in Emergency Management Information Systems. *Communication of the ACM* 50, 3 (March), 33–38.

Cullon, Charmayne. (1992). Human Factor Issues in the Nuclear Power Industry: Areas of Concern and Changes during the Eighties. *SAM Advanced Management Journal* (Winter), 16–20.

Han, Sung H., Yang, Huichul, Im, Dong-Gwan. (2007). Designing a Human-Computer Interface for a Process Control Room: A Case Study of a Steel Manufacturing Company. *International Journal of Industrial Ergonomics* 37, 5 (May), 383–93.

Jones, Ronald J. (1997). Corporate Ergonomics Program of a Large Poultry Processor. *American Industrial Hygiene Association Journal* 58, 2 (February), 132–37.

Lin, Y., and Zhang, W.J. (2005). A Function-Behavior-State Approach to Designing Human-Machine Interface for Nuclear Power Plant Operators. *IEEE Transactions on Nuclear Science* 52, 1 (February), 430–39.

Marklin, Richard W., and Wilzbacher, Jeremy R. (1999). Four Assessment Tools of Ergonomics Interventions: Case Study at an Electric Utility's Warehouse System. *American Industrial Hygiene Association Journal* 60, 6 (November–December), 777–84.

Moore, J. Steven, and Garg, Arun. (1997a). Participatory Ergonomics in a Red Meat Packing Plant, Part I: Evidence of Long-Term Effectiveness. *American Industrial Hygiene Association Journal* 58, 2 (February), 127–31.

Moore, J. Steven, and Garg, Arun. (1997b). Participatory Ergonomics in a Red Meat Packing Plant, Part II, Case Studies. *American Industrial Hygiene Association Journal* 58, 7 (July), 498–508.

Moss, T.H., and Sills, D.L. (eds.). (1981). *The Three Mile Island Nuclear Accident: Lessons and Implications*. New York: Academy of Sciences.

9 Maritime Application of Control Systems

Margareta Lützhöft and Monica Lundh

CONTENTS

9.1   INTRODUCTION TO SHIP CONTROL CENTRES

This chapter deals mainly with marine applications of control systems on board ships. However, over the past few decades, there has also been a development in the technology of management and control of seagoing traffic. This applies to particular phases of a sea voyage, for example, in the approach phases and when navigating narrow channels such as the English Channel and narrow straits such as the Malacca Strait. These types of control systems are increasingly becoming very similar to the traditional traffic control systems. In the past radio communications were used almost exclusively. Nowadays, these advance systems of traffic control and traffic management at sea will hopefully increase safety at sea. They may also change the role of pilots and harbourmasters, and their relations to the deck officers. Control centres and control rooms on board ships are obviously central to safety at sea. These are also a good example of more specialised forms of control centres. There are also similarities with developments in other areas of control rooms, in particular those applied to traffic control and traffic flow management in general.

9.2   BACKGROUND

The safe and effective operation of a ship depends upon a coherent set of control centres. These have evolved with shipping technology and manning; the radio compartment has come and gone, the engine control room grew and then changed focus from system control to system management, safety centres for passenger ships are still growing, cargo control has become more sophisticated, and the bridge has grown from a fairly simple space to a ship control centre filled with computing equipment.

9.2.1   THE BRIDGE

Research into bridge ergonomics and maritime human factor issues began in the 1950s. Earlier references to ergonomics (in the 1930s and 1940s) in trade journals and magazines are brief and infrequent and centre on visibility from the bridge and communications on and beyond the ship. In 1959 the British Ministry of Defence commissioned a study on integration of systems and layouts of bridges (Millar and

Clarke, 1978). A decade later Esso commissioned a study of merchant tanker bridges (Mayfield and Clarke, 1977; Clarke, 1978). The first substantive treatment of human factors and ergonomics on merchant ships' bridges seems to be a paper from 1971 (Wilkinson, 1971), which gives a thorough view on the evolution of bridges and bridge equipment, in particular from an ergonomic viewpoint. In Holland, human factors on the bridge have been considered and researched since the 1960s (Walraven and Lazet, 1964). In the 1970s there was a great deal of ergonomics research and development (Istance and Ivergård, 1978; Ivergård, 1976; Mayfield and Clarke, 1977; Proceedings of the Institute of Navigation National Maritime Meeting, 1977; Proceedings of the Symposium on the Design of Ships' Bridges, 1978).

At that time, ergonomists believed that maritime ergonomics had 'made a breakthrough' (Mayfield, personal communication). Unfortunately, this positive trend did not continue. According to an official at the Swedish Maritime Authority, at least Swedish ship-owners felt swamped by all the new regulations put out by the Swedish Maritime Authority. It was too much, came out too fast, and the development of maritime ergonomics more or less ground to a halt around 1980. To be fair, this was in part also due to other factors such as the issues not being on the checklist of requirements of purchasers. However, the necessity of considering ergonomics on board ship, in the context of technology, has been written about for at least 35 to 40 years. Unfortunately the emphasis still is on making humans adapt to computers and technology, whatever their limitations.

Research since then has been more limited. A major string of projects was Advanced Technology to Optimise Maritime Operational Safety (ATOMOS) I through ATOMOS IV, initially technological but swiftly changing to become human centred. Other research has been on the human aspects of the system such as bridge resource management (BRM) and fatigue. However, new technology has been introduced on the bridge at a fast pace. For instance, global maritime distress and safety systems (GMDSS) made the radio operator redundant but at the same time introduced distractions to the bridge watch-keeper (Sherwood-Jones et al., 2006), automatic identification systems (AIS) added more distractions and complexity (Blomberg, Lützhöft, and Nyce, 2005), and direct bridge control of more complex propulsion such as combined joysticks increased the mental workload of the operator. In sum, many systems have been added without a clear focus on the needs of the bridge watch-keeper, including some (perhaps too many) that do not ease the workload or simplify the task.

9.2.2   THE ENGINE CONTROL ROOM

In the engine room and engine control room, far less research has been done than on the bridge (Andersson and Lützhöft, 2007). However, new technology tends to be introduced on board at a rapid pace in the engine control room as well. This implies new demands on the knowledge and skills of the engineers, and thus creates a need for improvement of the crews' qualifications. In July 2005, due to an engine failure, one of the then largest container vessels, the *Savannah Express*, made heavy contact with a link span at Southampton Docks. One of the main reasons for the accident was that the ship's engineers did not have sufficient knowledge of the main engine control system or specific system engineering training to successfully diagnose faults. The subsequent report stated under the heading 'Training—General Conditions' that:

> Modern vessels increasingly rely on complex, integrated control and operating systems. Often these systems cannot be separated to enable operation of the equipment in a 'limp home' mode. The rapid introduction of such technology has placed ever-increasing demands on the shipboard engineers, who have often not had the requisite training with which to equip them to safely operate, maintain and fault find on this complex equipment.*

This statement gives clear evidence of how important the design of integrated alarm, monitoring, and control system is. Any interface with illogical, ambiguous, or cluttered design constitutes an increased risk and may promote errors.

On the positive side, in 2006, Liberia submitted a paper to the International Maritime Organisation's (IMO) Maritime Safety Committee (MSC), 'MSC 82/15/4 Role of the Human Element', which proposes a review of specific IMO instruments from an ergonomic perspective *inter alia* the guidelines for engine room layout, design, and arrangements described in 'MSC/Circ.834 Role of the Human Element' from 1998. The Maritime Safety Committee stated in 'MSC-MEPC.7/Circ.3 Framework for Consideration of Ergonomics and Work Environment' (2006) that a significant reduction of accidents to seafarers and human error can be obtained through the consideration of ergonomics and the working environment onboard ship. The framework considers five key areas on board ship:

- Manual valve operation, access, location, and orientation.
- Stairs, vertical ladders, ramps, walkways, and work platforms.
- Inspection and maintenance considerations.
- Working environment.
- The application of ergonomics to design.

The first three points are mostly relevant to the occupational risks to the engine room crew in modern ships. The fourth area (working environment) is universally relevant, since it encompasses also the psychological working environment and hence the general ability of the crew to work as a team. The last item, ergonomics is obviously, of particular interest in the current context. In essence, some promising initiatives have been taken in the area but much remains to be done. Later sections in this chapter strive to elaborate on the possibilities in that direction.

9.3   THE SITUATION TODAY

One example of work undertaken to address the issue of bridge ergonomics is the integrated bridge. The brief history of the integrated bridge system or integrated navigation system contains a few rather clear influences. A group of captains and pilots from Swedish and Finnish shipping companies saw the need for bridge technology

---

* Report on the Investigation of the Engine Failure of the *Savannah Express* and her Subsequent Contact with a Link Span at Southampton Docks, 19 July 2005. Maritime Accident Investigation Branch Report No: 8/2006 (2006).

that would help them navigate increasingly larger and faster vessels safely in the Swedish–Finnish archipelago. Since the 1960s, this community of practitioners and shipping companies have been cooperating with manufacturers. Their collaboration is partly driven by recognition of the complexity of the waters in which the ships sail, and because everyone recognised the value that new technology can have for sailing safely in difficult environments. Another influence towards the integrated bridge came from shipyards requiring manufacturers to deliver assembled solutions. The reasons were partly economical (cost savings on cabling) and partly for reasons of data and information integrity (systems needed to be interconnected). This does not mean that an assembled solution necessarily constitutes an integrated bridge, since the level of integration—from physical proximity of equipment to full-scale data-level integration—is mostly dependent on the purchaser's requirements. Finally, administrations from several countries were considering solo watch-keepers. This too demanded innovations on the bridge.

In subsequent decades the emphasis has been on designing an integrated naviga-tion system. By applying the concept of *integration work* to what humans do, we see that humans should be considered a part of these integration processes. Integration in this sense is about coordination, cooperation, and compromise. When humans and technology have to work together, the human (mostly) has to coordinate resources, cooperate with devices, and compromise between means and ends. What seafarers have to do to get their work done includes integration of representations of data and information and integration of human and machine work (Lützhöft, 2004). This is also made clear by Hederström and Gyldén (1992):

> When selecting the equipment, it must be borne in mind that it should be compatible with the other instruments to form an efficient work system as a whole. This is essential not only in the case of totally integrated systems, but also *if the integration is to be per-formed manually by the navigator.* (Hederström and Gyldén, 1992, 2; emphasis added)

The issue is also discussed by Courteney (1996) who vividly describes how the captain of a ship is torn between stakeholders and has to provide coherence and balance to the situation.

Several anecdotes from the maritime domain describe how the placement of a piece of equipment is decided not by standards, rules, or human-factor guidelines but by the length of electric cable available to the installer when needed. One ship, built as a container ship in the 1960s, rebuilt as a cruise liner in 1990, and visited by the first author in 2001, had fifteen different manufacturers' names on the navigation, control, and communication equipment. This in itself leads to many inconsistencies for the seafarers to overcome. Furthermore, it is common that when new equipment is installed, in many cases the older equipment is left where it is. Apart from creating a cluttered and perhaps nonoptimal layout, this also entails extra work, such as finding, choosing, and evaluating which equipment to use. A related issue is the use of individ-ual audible alarms in most equipment—the beeps emitted by the various instruments sound very similar, and finding the source of the alarm is often complicated. Resolv-ing these technical issues does not resolve matters completely. Adding to the complex-ity of operations are issues such as new legislation and crew turnover. Seafarers have to perform integration work, since technological resources are never constant.

Current incident reports still show too many collisions and groundings. The problems at 'the sharp end' can be very basic, for example, when a single watch-keeper falls asleep. Furthermore, the reality of current bridges is that many basic aspects of control room ergonomics are not addressed, as for example found in the MTO-sea project (ongoing 2007). As an illustration, basic console design is an area for which there is ample guidance but which is not always used. An example is current consoles that are not always designed so that the operator can stand and operate the console or sit and look out of the window. Proper sit–stand designs have been in the standards for a long time but are not understood or widely implemented. An exception is shown later in this chapter.

Colloquially, the engine room is often described as the heart of the ship whereas the bridge is seen as the brain. However, neither one can perform anything worthwhile without the other. It cannot be emphasised enough that the engine control room and the bridge (and many more workplaces on board a ship) should not be optimised separately. By this we mean that the common goals of both should be considered, not that they both should be designed in the same way. Planning, organisation, and design must be performed throughout with common goals in mind: to ensure communication and a common awareness of what is happening, and ultimately to move cargo or passengers from one point to another, while considering the safety of crew, passengers, and cargo, as well as of the ship and the environment. Today, maritime security is an added concern. In addition, shipping is, of course, a business that should be successful from a financial viewpoint. To perform this task, the ship must be viewed as a system where all parts contribute to the fulfilment of the task.

9.3.1   THE WAY AHEAD: UNDERSTANDING EACH OTHER

In order to understand the demands and requirements made by both departments (that is, the engine room and the bridge), it is essential to establish common knowledge and an understanding of each other's working procedures. It is vital that the checklists that are to be used in abnormal conditions are discussed by representatives from both departments, for example, the routines to be followed if the main engine fails to manoeuvre. The engine department is to be informed whenever the situation demands a manned engine control room—for instance, in narrow waterways or when approaching or leaving harbour. It is also necessary to increase the awareness of the situation for the engine room crew. They can benefit greatly from knowing what is happening on the bridge and perhaps even being able to track the ship's position. A cheap and effective solution to increase such awareness may be the addition of an outside, forward-looking camera to the closed-circuit television (CCTV) system, accessible also from the engine control room. Of course, the engine room crew must be trained to perform the correct actions in different situations and the bridge must be aware of the approximate time needed to perform these correct actions. If the acquired approximate time is known, the planning of the right action—for instance, to drop the anchor—can be facilitated and accidents can be avoided. This puts demands on the design of the equipment in the engine room. All emergency running of any equipment must have easy access. As such, it is suboptimal to hide important functions behind endless menus in a complicated computer-based structure.

During the operation of ships, especially tankers, certain operations are vital for the safety of the crew and the ship. One example is the production and distribution of inert gas. To be able to produce enough inert gas of good quality, the boiler must be operated at high load. This implies that, for instance, turbo generators must be run at heavy load. Whenever a cargo tank is empty, the cargo pumps reduce their energy consumption, less steam is required, the boiler decreases its capacity, and the quality of the inert gas deteriorates (lower load equals an increase in oxygen content). This could eventually result in an explosive atmosphere in the tanks if the appropriate actions are not taken. Such a situation demands good communication between the officer in charge on deck and the officer in charge in the engine room. Any unwanted effects caused by any changes in the requirement of steam can be monitored and avoided if the engineer is aware and prepared for these changes. Thus, the design of the engine room must enable an easy and reliable means of communication with the bridge and the navigating officers.

9.3.2 DESIGN OF SHIP CONTROL CENTRES

It is important to take into consideration the influence of some domain-specific contextual factors on the ship control centres (SCC):

- The size and type of ship.
- Crew composition (number of operators, sometimes including a pilot).
- For specialised ships intended for specific trades, design could also be influenced by, e.g., the:
  - Predominant weather in the area
  - Layout of ports used frequently
  - Type, nature, and specific requirements of the cargo

In a later section we discuss a few ship-specific design aspects, but many aspects of the design of ship control centres are similar to other control rooms. Most essentially, operators have to be able to reach and see all the information that they need to perform their tasks. This kind of general information and control room design is well covered elsewhere, and also in some ship-specific detail, in standards and guidelines (IMO, 2000; ISO, 2007).

A brief discussion is included here, which discusses the importance of human factors engineering (HFE) for ship control centres, and which is based on the ATO-MOS projects. HFE in this context means the comprehensive integration of human characteristics into the definition, design development, and evaluation of the SCC to optimise human-machine performance under specified conditions. Task and information analysis of SCC operations should be used to design each application and the required data exchange with other applications, considering that SCC applications software may be provided to support all ship operations. Examples are: navigation, propulsion, communications, manoeuvres, maintenance, alarm-monitoring and control systems, plus, in some cases, specialised applications such as ship administration, cargo management, and hull stress monitoring.

The size and type of the ship may restrict the space available for the SCC. Since the SCC is manned continuously, the effect of layout, decoration, and design on crew well-being should be considered. The operability of the SCC in both routine and emergency operations should be considered. Operability factors include: having all instruments at hand, having working space sufficient to allow for easy movement, having ergonomically designed equipment, and having adequate and appropriate surroundings. It should furthermore be considered that the design for increased interaction by communication or sharing of information may give operational advantages in both routine and emergency operations.

9.3.3 REGULATORY SUPPORT FOR USABILITY

9.3.3.1 The Bridge

For the bridge, there exist several documents to support design:

- SOLAS (International Convention for the Safety of Life at Sea) Chapter V/15 contains seven aims for how the work and workplace on the bridge should be designed (IMO, 1974), and these aims will be discussed in Section 9.4, 'Design of the Bridge'.
- A comprehensive standard, International Organisation for Standardisation (ISO) 8468, contains an abundance of guidelines and recommendations on bridge layout and associated equipment.
- IMO provides MSC/Circ.982 (IMO, 2000) which contains guidelines on ergonomic criteria for bridge equipment and layout.*
- An International Association of Classification Societies (IACS) Recommendation (IACS, 2007) is also available and discusses in more detail how to meet the aims addressed by SOLAS regulation V/15. The requirements in this document cover SOLAS regulations and applicable parts of MSC/Circ.982 in order for the standard to be useable as a stand-alone document for the purpose of approval work during the building process. The document discusses the areas addressed by SOLAS regulation V/15: bridge design, design and arrangement of navigational systems and equipment, and bridge procedures.
- The marine sector uses its own ISO TC8 standards, but adoption of the IC159 ergonomics standards has still to be achieved.

Furthermore, many classification societies provide guidelines and certification of various aspects of ergonomic and human factor issues. One must, however, be aware that using guidelines to optimise the usability of many separate instruments does not necessarily mean that usability is ensured for the system or SCC as a whole. For this, a usability process should be used throughout.

9.3.3.2 Engine Control Room

Contrary to the situation of bridges, there is no regulatory support which completely covers the layout, design, and arrangement of the engine department. As mentioned

---

* Available at http://www.imo.org/includes/blastDataOnly.asp/data_id%3D1878/982.pdf.

earlier, IMO MSC/Circ.834 gives only a brief description of the nonmandatory guidelines for engine-room layout, design, and arrangement.

The International Organisation for Standardisation (ISO) has developed a standard, ISO 11064, which gives advice on the ergonomic design of control centres. Other ISO standards which might be of interest are for instance:

- ISO 9241 'Ergonomics of human-system interaction—Part 400: Principles and requirements for physical input devices'.
- ISO 9355-3:2006, 'Ergonomic requirements for the design of displays and control actuators—Part 3: Control actuators'.
- ISO 9241-110:2006, 'Ergonomics of human-system interaction—Part 110: Dialogue principles'.

Some classification societies also offer frameworks to support the construction of engine rooms and engine control rooms but this lack of a uniting set of rules and regulations implies difficulties to provide an overview of this extensive area. Much more research is needed, and this need is currently being addressed in a project at Chalmers University of Technology, Gothenburg, Sweden.

9.4   DESIGN OF THE BRIDGE

SOLAS V/15 (hereafter used interchangeably with 'the Regulation') considers ship control as a sociotechnical system (see www.imo.org). The provision of information on the bridge and use of information by the bridge team and pilot are covered equally in the Regulation, which is concerned with resource management rather than just equipment design or training. There are a number of factors that need to be considered when making a decision so that the aims can be achieved. These factors are based on established human factors and safety findings. The factors to be taken into account when making a decision are:

- Manning operations and procedures
- Training
- Equipment and system design
- Bridge layout

In addition to the above, one should, for specialised ships built for specific purposes, consider any characteristics that may be relevant for the way the bridge is used. Although we mainly deal with equipment and system design and bridge layout here, it is important to consider the other aspects above to maintain a systems view. The Regulation requires that decisions be made with the intention of meeting the seven aims. These aims are mostly concerned with the performance of the work system, although there are aspects concerned with product characteristics. Table 9.1 (from the ATOMOS IV project) indicates this.

The development process characteristics column in Table 9.1 is empty, which means that SOLAS V/15 does not specifically have any requirements for the development process itself but rather for the results of such a process.

TABLE 9.1

Required Characteristics of SOLAS Regulation V/15

Aim Number	Work System Performance Characteristics	Product Characteristics	Development Process Characteristics
1.1	Facilitating the tasks to be performed by the bridge team and the pilot in making full appraisal of the situation and in navigating the ship safely under all operational conditions		
1.2	Promoting effective and safe bridge resource management		
1.3	Enabling the bridge team and the pilot to have convenient and continuous access to essential information that is presented in a clear and unambiguous manner	Using standardised symbols and coding systems for controls and displays	
1.4		Indicating the operational status of automated functions and integrated components, systems and/or subsystems	
1.5	Allowing for expeditious, continuous and effective information processing and decision making by the bridge team and the pilot		
1.6	Preventing or minimising excessive or unnecessary work and any conditions or distractions on the bridge that may cause fatigue or interfere with the vigilance of the bridge team and the pilot		
1.7	Minimising the risk of human error and detecting such error if it occurs, through monitoring and alarm systems, in time for the bridge team and the pilot to take appropriate action.		

Source: ATOMOS IV Project.

The ATOMOS projects (I to IV) studied, among many other issues, the process of making a good design. One result was the design and redesign templates to be used when designing a new bridge or making changes to an existing bridge. The achievement of work system performance characteristics can be assessed only at a late stage of implementation. However, it is possible to provide guidance on product characteristics and process characteristics to support the process of moving from a decision to make a change through to its implementation. These characteristics are the factors that will influence the achieved work system performance, and it is possible to set ergonomic criteria for these factors to reduce the risk of the implementation of the change. As shown in Table 9.2, most factors affect the achievement of most aims. The original table also contains guidance on the impact of ergonomic criteria on training, manning, and operations. Here, we show only the design-related criteria.

9.4.1 Specific Design Aspects

A few aspects are unique for ships and are therefore discussed here. For instance, operators could in special cases need a console at which work can be performed both sitting down and standing up, in order for the bridge personnel to keep a good look-out; thus, special consideration must be given to console and chair design. Alternatively, on some ships, typically high-speed craft, all work can and must be performed sitting down, necessitated by the often violent motions of the vessel in bad weather.

Another consideration is that work on the bridge is 24 hours a day, 7 days a week, daytime and night-time. Thus the displays and instruments must afford the dual function of being clearly readable in bright sunlight while also not disturbing night vision during darkness. The number of operators can vary from one or two, to (at times) up to five or more. During normal sea voyages the bridge crew may consist of one nautical officer and at times of darkness or low visibility, a lookout. When manoeuvring in restricted waters or port areas, the bridge has to accommodate at least a master, a pilot, perhaps two nautical officers, and a helmsman. This places high demands on both workplace design and procedures to make sure all team members can access relevant information and necessary controls at all times.

Even though ergonomic guidelines exist, they should always be considered in context. It is crucial to know about the functions to be performed in the ship control centres. An example of a bridge workplace is shown in Figure 9.1. This probably works well when manoeuvring normally and looking forward, but we see how it does not work well when manoeuvring close to a towering oil rig. One person manoeuvres the ship while the other is keeping a lookout. Ergonomics of both working positions are less than perfect. The position and posture of the lookout is evidently poor; the controller has an overextended elbow and obviously cannot see all he needs to see.

9.4.2 Illumination and Lighting

A satisfactory level of lighting should be available to enable the bridge personnel to complete such tasks as maintenance, chart, and office work satisfactorily, both at sea and in port, daytime and night-time (IMO, 2000). The following bulleted list and table contain an overview of important lighting points (IMO, 2000):

- Red or filtered white light should be used to maintain dark adaptation whenever possible in areas or on items of equipment requiring illumination in the operational mode. This should include devices in the bridge wings.
- High contrast in luminance between the work area and surroundings should be avoided, i.e., luminance of the task area should not be greater than three times the average luminance of the surrounding area.

TABLE 9.2
Relationship between the Aims in SOLAS Regulation V/15 and Factors Used in the Templates for Submission Statements

Factors	Aim from SOLAS Regulation V/15						
	1.1 Facilitating the tasks to be performed by the bridge team and the pilot in making full appraisal of the situation and in navigating the ship safely under all operational conditions.	1.2 Promoting effective and safe bridge resource management.	1.3 Enabling the bridge team and the pilot to have convenient and continuous access to essential information that is presented in a clear and unambiguous manner, using standardised symbols and coding systems for controls and displays.	1.4 Indicating the operational status of automated functions and integrated components, systems and/or subsystems.	1.5 Allowing for expeditious, continuous, and effective information processing and decision making by the bridge team and the pilot.	1.6 Preventing or minimising excessive or unnecessary work and any conditions or distractions on the bridge that may cause fatigue or interfere with the vigilance of the bridge team and the pilot.	1.7 Minimising the risk of human error and detecting such error if it occurs, through monitoring and alarm systems, in time for the bridge team and the pilot to take appropriate action.
Aims							
Equipment selection and design	Equipment will need to facilitate the operators' tasks. Equipment will need to support operation in all conditions.	The equipment needs to be able to support BRM, e.g. SIC, oversight and supervision, cross-checking.	User interfaces will need to be consistent and well designed. They may also need to be dependable to meet the continuous access requirement. Essential information will need to be identified as such. Cross-system consistency will be required.	FMECA etc. will need to be undertaken to ascertain the operational status of equipment and to determine what indication should be provided.	Equipment needs to be able to support human information processing, decision making, and teamworking.	Design and selection needs to avoid unnecessary work, e.g., software menus. The design should not cause distraction (loss of dark adaptation, irrelevant messages).	Design needs to address human error prevention and recovery, including alarm philosophy for acceptance, inhibit, transfer, etc.

| SCC layout and design | Information provision to each operating position will need to be fully considered. | The layout needs to allow supervision and teamwork. | SCC design needs to proceed in an integrated manner. Proving that this aim has been met will require user-centred design through the life cycle. | Some system integration may be required to ensure that operational status is defined, assessed, and indicated. | Layout needs to provide good access to information and to allow teamworking. | Equipment location needs to avoid causing distraction. If it is not part of SCC operations it should not be on the bridge. If it is relevant, then it needs to be appropriately located. | Alarm messages need to support the bridge team and pilot in location and visibility. |

Source: ATOMOS.

FIGURE 9.1   Manoeuvring a ship close to an oil rig cannot be done by one person. (Drawing by E. Jacobson after a photo. With permission.)

- The lighting system should enable the bridge personnel to adjust the lighting in brightness and direction as required in different areas of the bridge and by the needs of individual devices. Table 9.3 lists the recommended general illumination levels and colours.
- A light-dimming capability should be provided.
- Utmost care should be taken to avoid glare and stray image reflections in the bridge environment.
- Lighting sources should be designed and located to avoid creating glare from working and display surfaces.
- Reflection in windows of devices, instruments, consoles, and other reflective enclosures should be avoided.
- Devices should be designed and fitted to minimise glare, reflection, or obscuration by strong light.

Related to lighting is the choice of colours for interior surfaces. For the interior, unsaturated colours should be chosen that give a calm overall impression and minimise reflectance. Bright colours should not be used. Dark or mid-green colours are recommended; alternatively blue or brown may be used. In spite of these guidelines, a very common complaint is about the lack of dimming (Lützhöft et al., 2007). The bright light destroys night vision, disturbs the keeping of a safe lookout, and distracts attention (see Figure 9.2).

TABLE 9.3
General Illumination for the Bridge

Place	Colour/Illumination
Bridge, night	Red or filtered white, continuously variable from 0 to 20 lux
Adjacent corridors and rooms, day	White, continuously variable from 0 to at least 300 lux
Adjacent corridors and rooms, night	Red or filtered white, continuously variable from 0 to 20 lux
Obstacles, night	Red spotlights, continuously variable from 0 to 20 lux
Chart table, day	White floodlight, continuously variable from 0 to 1000 lux
Chart table, night	Filtered white floodlight or spotlights, continuously variable from 0 to 20 lux

Source: IMO (2000).

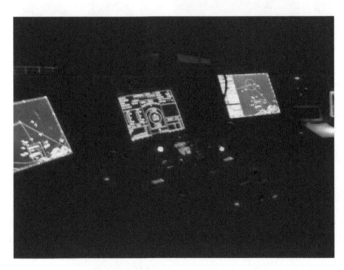

FIGURE 9.2 Many modern displays cannot be dimmed enough and disturb night vision. (Photograph from MTO-Sea Project. With permission.)

Not only modern displays cause dimming trouble. Many disturbing points of light are visible on many bridges, and the problem mainly stems from the amount of different manufacturers providing equipment, and the lack of integrated or system design thinking. Homemade dimmers abound, some more innovative than others (see Figure 9.3).

9.4.3 MOVING AROUND

Another aspect to be considered is that this workplace moves. Therefore, to avoid slips, trips, and falls, the wheelhouse, bridge wings, and upper bridge decks should have nonslip surfaces. Furthermore, there should be no sharp edges or protuberances that could cause injury to personnel, and sufficient hand- or grab-rails should be

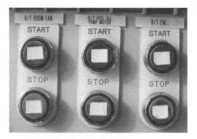

FIGURE 9.3 Examples of homemade dimmers. (Photographs from MTO-Sea Project. With permission.)

fitted to enable personnel to move or stand safely in bad weather when the vessel not only moves but is liable to list, pitch, and roll. Protection of stairway openings should be given special consideration. Attention should be paid to bridge wing doors, when present, to ensure they can be opened safely and easily.

9.4.4 BRIDGE LAYOUT

A number of basic functions describe the work on a ship's bridge. A function can be a group of tasks, duties, and responsibilities necessary for the operation of the ship and carried out on the bridge. The most basic functions which have to be taken into account when designing this workplace are (IMO, 2000):

- Navigating and manoeuvring
- Monitoring
- Manual steering
- Docking/undocking
- Planning and documentation
- Safety
- Communication

General guidelines exist for the layout of bridges. These apply both to larger ships, which often nowadays use a so-called cockpit bridge, and to smaller ships which use a more traditional layout. The workstations for navigating and manoeuvring, monitoring, and for the bridge wings should be planned, designed, and placed within an area spacious enough for not fewer than two operators but close enough for the workstations to be operated by one person. As the name suggests, a cockpit bridge is based on the cockpit of an aeroplane. All functions and controls needed by

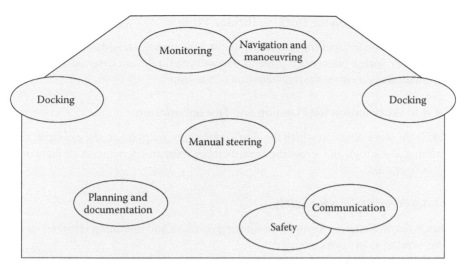

FIGURE 9.4   Bridge layout with suggested work stations and their functions. (Adapted from IMO, 2000. With permission.)

a single operator or a team of operators are placed centrally and gathered around the operators. An example is shown in Figure 9.4.

Here, the functions listed above are used to describe the workstations. More details of what to include and where to the instruments and controls are found in IMO, ISO, and IACS documents. The specific design and placement of individual instruments and controls will always depend on ship-specific factors such as trade, ship size, and crew size, and will not be discussed here in detail. A brief explanation of the functions (IMO, 2000) follows.

9.4.4.1 Workstation for Navigating and Manoeuvring

The main workstation for a ship's handling consists of crew members working in seated-standing position with optimum visibility and integrated presentation of information and operating equipment to control and consider ship's movement. From this position it should be possible for a crew member to operate the ship safely, in particular when a fast sequence of actions is required.

9.4.4.2 Workstation for Monitoring

The workstation from which the crew member is operating equipment and from which his or her surrounding environment can be permanently observed should be in a seated-standing position. When several crew members are working on the bridge, it serves for relieving the navigator at the workstation from navigating, manoeuvring, or carrying out control and advisory functions by master or pilot.

9.4.4.3 Workstation for Manual Steering (the Helmsman's Workstation)

The workstation from which the ship can be steered by a helmsman, as far as legally or otherwise required or deemed to be necessary, is preferably conceived for working in seated position.

9.4.4.4 Workstation for Docking (Bridge Wing)

The workstation for docking operations on the bridge wing should enable the navigator together with a pilot (when present) to observe all relevant external and internal information and control the manoeuvring of the ship.

9.4.4.5 Workstation for Planning and Documentation

This is the workstation at which the ship's operations are planned (for example, route planning, deck log), and where the crew member fixes and documents all facts of the ship's operation.

9.4.4.6 Workstation for Safety

This is the workstation at which monitoring displays and operating elements or systems serving safety are colocated.

9.4.4.7 Workstation for Communication

This is the workstation for operation and control of equipment for distress and safety communications (GMDSS) and general communications (IMO, 2000).

These functions are intended as high-level descriptions and their specific placement (suggested in Figure 9.4) can be discussed. It is also an open question whether separating functions in this way gives good usability with regard to the entire system. However, the figure should be seen as a guide to what work is to be performed on the bridge. It is not a design guide and does not imply that these functions must be separated from each other. It is also conceivable that all functions should be placed together, gathered near the operator. This will depend on several issues that should be discussed by the shipping industry. For instance: Are all these functions really the responsibility of the watch-keeper? Should they all be performed on the bridge? Who will be working on the bridge, and what competence is needed in the future?

For smaller ships, a different layout may be relevant. It could, for example, be due to cost or space limitations. Figure 9.5 is a suggestion prepared by IACS (2007) for ships smaller than 3000 gross tonnes.* The layout is traditional with consoles across the breadth of the bridge.

When using this layout, careful thought must be put into placement of aids and instruments, and adopting a functional view of the work stations. Otherwise an operator may have difficulty performing tasks while on watch. Figure 9.6 and Figure 9.7 show photographs of such a bridge on a large ship. Many instruments and aids are too far from each other, and the long console does not afford good teamwork. One example of this is the inability to reach the VHF radio handset while looking at the radar. The same is true for reaching engine controls when working at the radar. On board ship, each of these situations is quite conceivable and commonplace. The operator may want to talk to another ship or shore station to confirm the position of something or someone visible on the radar screen. The second situation may also

---

* 'Gross tonnage' is a way of describing ship size and refers to the volume of all the ship's enclosed spaces.

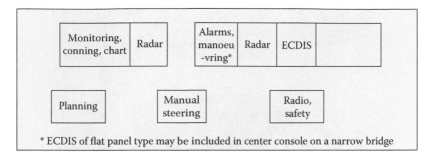

* ECDIS of flat panel type may be included in center console on a narrow bridge

FIGURE 9.5 Bridge layout for ships of less than 3000 gross tonnes. (Adapted from IACS, 2007. With permission.)

FIGURE 9.6 Traditional long console bridge, view from the starboard to port side. (Personal photograph of M. Lützhöft.)

occur, when avoiding close-quarter situations with other ships or manoeuvring in restricted waters.

More specific examples of interaction problems and design flaws that could lead to risks can be found in the smallest things. Post-Its, notes, and instruction often give clues. In Figure 9.8 we see a striking example. To fully understand the implications, we must understand the context. In this specific example one can easily imagine the consequences of not performing this action (start pump), firstly leading to a mechanical failure (due to overheating), costly in itself, and secondly in the worst-case scenario leading to a grounding or collision after the bow thruster fails (Lützhöft et al., 2007).

One function that is missing from the IMO list but that is performed to an increasing degree on the bridge is administrative and office duties. It is not safety critical but could conceivably be included in the planning and documentation function. The

FIGURE 9.7 Traditional long console bridge, view from the port to starboard side. (Personal photograph of M. Lundh.)

FIGURE 9.8 Examples of instructions that indicate interaction problems. (Photograph from MTO-Sea Project. With permisssion.)

inclusion of such a function should, however, be considered very carefully, since it could add disturbance to the bridge operation and reduce vigilance. Hence it would be in contradiction to the sixth aim of SOLAS V/15. It is an issue to discuss: who performs it and where it is to be done.

Other distractions that from time to time are found on bridges should be discussed critically and, if found inappropriate, removed. The prime example is mobile phones. Private phones are questionable in themselves, but also ship phones need to be considered. If not removed, a policy should be discussed for calling in to (and out from) the ship, and who answers at what times. It is not uncommon for the phone to ring at times when the bridge operators need to focus on their primary tasks. Further distractions may be private MP3 players or similar devices, and also radios, TVs, and even sofas where often meetings are held by those not on watch.

Figure 9.9 shows a good and well-thought-through bridge solution. In this instance, an in-house ergonomist was engaged in the design process together with the manufacturer, and prospective users were involved throughout, taking into account lessons learned from other ships, adapting the ships to their particular trade and the functions to be performed. Special features to note are the consoles where keyboards are placed. A problem has always been to design a workplace that functions well for work both standing up and sitting down. Here, this is solved by installing tables that can be raised and lowered, like the ones now available for offices. We may note

FIGURE 9.9   A modern bridge with fixtures and fittings that can be adapted to different needs. (Design by Wallenius Marine AB/Furuno Finland. Photograph used with permission.)

also that there is available space on the left side of the console where new equipment can be installed, if necessary. As mentioned, the evolution of the bridge and new technological aids is an ongoing process, which can lead to difficulties if the bridge becomes 'full', and new aids may get suboptimal placement. However, this does not mean that leaving a space affords optimum placement for future equipment but only that the designers at least took the issue into account.

Another issue on these ships, and many others, has been that when berthing, the captain wishes to see the side of the ship and the berth. At the same time he needs to control the ship's movements. Earlier the captain would have to stand with his back to the controls to see the side of the ship and the berth. Now, the console is tilted upwards, good ergonomics is incorporated, and the work has been made easier (see Figure 9.10 and Figure 9.11).

Even with a good process for including human factors into the design, some unexpected changes may well have to be made after the fact. One example is a ship designed much like the above. They were trading on East Asian ports, and it became obvious that the local pilots wished to stand close to the windows. Although the bridge and consoles were designed to allow this, from this position they could not reach the VHF radio handset. As the VHF is much used in piloting waters, an additional VHF radio handset had to be installed next to the window, in front of the main console.

9.5   DESIGN OF THE ENGINE CONTROL ROOM

As demonstrated previously, there are not many legal requirements to the design of the engine control room (ECR) and little guidance for this control centre. Thus the discussion in this section will be of a more informal character and based on

FIGURE 9.10   View towards the bridge wing with new console. (Design by Wallenius Marine AB/Furuno Finland. Photograph used with permission.)

FIGURE 9.11   New bridge wing console, adapted to work when berthing. (Design by Wallenius Marine AB/Furuno Finland. Photograph used with permission.)

empirical data (Andersson and Lützhöft, 2007). As an example of the possibilities to use bridge guidelines for the ECR, we here present Table 9.4, which shows that if SOLAS V/15 were to be adapted to the engine department, its contents need little adjustment (compare Table 9.4 to Table 9.1).

We continue with a discussion of problem areas in the design of the ECR, supported by illustrations. The central piece of equipment here is, of course, the alarm system. The information provided by the alarm system must be easily accessible

TABLE 9.4
Suggested Adaptation of SOLAS V/15 to the Engine Control Room

Aim Number	Adaptation
1.1	Facilitating the tasks to be performed by the ECR team in making full appraisal to the situation and in running the ship safely under all operational conditions
1.2	Promoting effective and safe ECR resource management
1.3	Enabling the ECR team to have convenient and continuous access to essential information that is presented in a clear and unambiguous manner
1.4	Indicating the operational status of automated functions and integrated components, systems, and/or subsystems and alarms
1.5	Allowing for expeditious, continuous, and effective information processing and decision making by the ECR team
1.6	Preventing or minimising excessive or unnecessary work and any conditions or distractions in the ECR that might cause fatigue or interfere with the vigilance of the ECR team
1.7	Minimise the risk of human error and detecting such error if it occurs, through monitoring and alarm systems, in time for the ECR team to take appropriate action

from every position in the control room. This is also valid for the acknowledgement of alarms. The development of technology has led to an increase in the number of alarms on board. Alarms will be further discussed in a later section.

In order for an operator to maintain and update his or her mental model of the monitored processes both on the bridge and in the ECR, analogue presentations are often more helpful than digital. Reading errors are more common on digital instruments than analogue instruments. Analogue gauge indicators are in most cases preferred by engineers to digital gauge indicators (see Figure 9.12 and Figure 9.13). A specific value is more difficult to remember than the position of an indicator on an instrument. If the analogue instruments are designed in a way that implies that the needles of the instruments have the same position under 'normal operation', any deviations from the normal are easily detected (see Figure 9.14) (Wagner, 1994).

The instruments also need to be arranged due to their affiliation, the so-called functional grouping (see Figure 9.15). The layout must clearly show which knobs, switches, and indicator lights belong to a certain piece of equipment. For instance, controls for a pair of pumps, starboard and port, must be consistently mapped on the control panel in the same way (starboard and port). However, due consideration must be given to whether the control room or display surface is facing forward or aft. An issue discussed in many control rooms, as well as the ECR, is the loss of mimic boards, which often are replaced with computerised overviews. Care should be taken to provide operators with good overviews of systems, as well as giving them possibilities to zoom in on details.

FIGURE 9.12   Analogue instruments. (Personal photograph of M. Lundh.)

FIGURE 9.13   Digital instruments. (Personal photograph of M. Lundh.)

9.5.1   Engine Control Room Layout

The design of the engine control room should be guided by the functions that are to be performed there. A number of basic functions describe the work in a ship's ECR and can be described as a group of tasks, duties, and responsibilities necessary for the operation of the propulsion of the ship and carried out in the ECR. The most basic functions that should be taken into account when designing this workplace are:

- Normal operation
- Monitoring

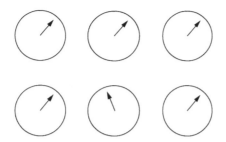

FIGURE 9.14 Deviation from the 'normal'. (Image from E. Wagner. With permission.)

FIGURE 9.15 Functional grouping of instruments. (Photograph from E. Wagner. With permission.)

- Emergency operation
- Planning and documentation
- Administrative duties
- Safety
- Communication
- Stand-by area

These functions are based on the functions described by IMO (2000) for the bridge, and are to be seen as a suggestion, open to discussion and revision. The specific design or placement of instruments is not discussed here, as these are dependent on the ship's type and size. As far as ergonomics is concerned, relevant guidance for design is available in order to place the equipment within reach and sight. The workstation must also be adapted for both standing and seated work positions, as applicable. This is so that the working position can be varied during long work shifts.

Figure 9.16 shows a suggestion of an ECR layout, taking into account the above functions. The division of functions on these figures does not imply that they must or should be separated from each other physically. Figure 9.17 shows a good example of a well-designed ECR, where engineers are placed centrally with a good view over system and alarm status.

9.5.1.1 Workstation for Normal Operation

This is the main workstation for the operation of the engine room and its auxiliary equipment.

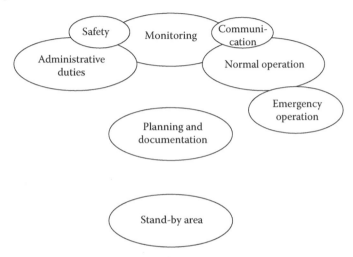

FIGURE 9.16   Suggested layout of ECR, with basic functions.

FIGURE 9.17   A traditional design of an ECR using analogue techniques. The on-duty engineers are located centrally and have a good view of the operational status and the alarm status. (Personal photograph of M. Lundh.)

9.5.1.2 Workstation for Monitoring

This is the workstation from which process values and alarms can be monitored.

9.5.1.3 Workstation for Emergency Operations

This is the workstation from which emergency running of the equipment can be performed.

9.5.1.4 Workstation for Planning and Documentation

This is the workstation at which the operation and maintenance of the engine room and engine control room are planned.

9.5.1.5 Workstation for Administrative Duties

This is the workstation at which members of the crew do administrative tasks such as e-mail and correspondence, the planning of maintenance work, inventory of available spare parts, and ordering of spare parts. If at all possible, taking into account manning restrictions, these duties should be performed at another location away from the control room.

9.5.1.6 Workstation for Safety

This is the workstation at which all safety-related duties can be performed. These include such duties as the operation of fire pumps and monitoring of the fire alarm systems.

9.5.1.7 Workstation for Communication

This is the workstation from which communication to the bridge as well as external communication can be performed.

9.5.1.8 Workstation for the Stand-by Area

This is the workstation where waiting time can be used.

9.5.2 Small Engine Control Rooms

The greatest challenge is, of course, the design of small ECRs. The smaller the available space, the more thought must be spent in the design. If not, the work in the ECR risks becoming restrained through overcrowding (if more than one crew member is present) or inefficient (if a solitary crew member cannot reach all the controls and equipment needed). Figure 9.18 and Figure 9.19 show a suggested layout for the smaller ECR, again fitting in the basic functions.

The tasks that are performed in the ECR have changed in recent years and nowadays engineers have to perform more computer-based tasks than before. If the crew is so few in number that administrative work must be performed during the control room watch, the layout of the ECR must take this into consideration and provide a desk. However, one should be aware that this may contradict the suggestions in

FIGURE 9.18   A narrow ECR where passage is restricted by desk chairs. (Personal photograph of M. Lundh.)

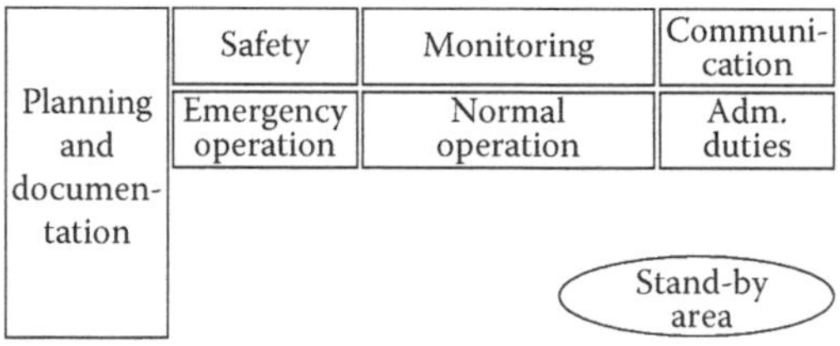

Planning and documentation	Safety	Monitoring	Communication
	Emergency operation	Normal operation	Adm. duties

Stand-by area

FIGURE 9.19   Suggested layout for a small ECR, with basic functions.

Table 9.4, aim 1.6: 'Preventing or minimising excessive or unnecessary work and any conditions or distractions in the ECR which might cause fatigue or interfere with the vigilance of the ECR team'.

If it is still deemed necessary, the layout should allow administrative work to be performed while the engineer still has a good overview of process and alarm information and can reach the acknowledgement button to the alarm system. Figure 9.20 shows an example of the opposite.

FIGURE 9.20   A narrow desk in the ECR which is in a corner. This leaves the engineer with his back to the operating panels and the alarm system. (Drawing by E. Jacobson from a photograph. With permission.)

It is also important to have a briefing area where meetings with the engine room crew can take place. The ECR is the area where discussions and sharing of information can take place without risk of misunderstandings due to the noise level in the engine room. The engine crew also need a stand-by area where they can relax and take a break from the noisy and hot environment, which in many cases the engine room represents (see Figure 9.21). The area can, for instance, also be used by a stand-by engineer waiting for a departure or waiting in a lock.

9.6   ALARMS

Alarms and alarm management are a growing problem, both on the bridge and in the ECR. Many alarms are necessary on the bridge, to indicate the status of safety-critical equipment or to alert the operator regarding a dangerous situation. However, the number of alarms and indications has increased from solely being related to navigation to include engine room/platform systems and now also communication/radio station alarms. Many instruments have their own alarm, and systems and alarms are not integrated. Many alarms are statutory requirements, but that does not make them appropriate all the time. As if this were not enough, much equipment designed for land and office use is making its way on to ships. Faxes, printers, and computers on the bridge all add to the plethora of alarms, making it even more difficult to judge

FIGURE 9.21   Stand-by area in close vicinity of the operating area. (Personal photograph of M. Lundh.)

where they come from (digital beeps are hard to place in space). It is also very difficult to judge the degree of urgency. In the ECR, we also see and hear more alarms, and alarm systems are more complex than before. The presentation of alarms (for example, long alarm lists on paper or screen) often makes it hard to interpret and prioritise the alarms (see Figure 9.22). Especially in emergency conditions, it would be beneficial if the alarm presentation helped operators pinpoint the initiating factors. If it were possible it would be helpful to suppress certain less urgent, secondary, alarms.

IMO 982 provides some guidance for alarm management and alarm design, but it does not solve all problems. The following list is based on IMO (2000), with some added comments, and in most cases applies to both the bridge and the engine control room.

9.6.1   Alarm Management and Design

- A method of acknowledging all alarms (i.e., to silence audible alarms and to set visual alarms to steady state), including the indication of the source of the alarm, should be provided at the navigating and manoeuvring workstation, to avoid distraction by alarms that require attention but have no direct influence on the safe navigation of the ship and that do not require immediate action to restore or maintain the safe navigation of the ship. In order to judge this, alarm information must be available at the same place.
- The alarm indicators and controls of the fire alarm and emergency alarm should be located at the safety workstation.
- Alarms should be provided to indicate failure or reduction in the power supply that would affect the safe operation of the equipment.
- Alarms should be provided to indicate sensor input failure or absence.
- Alarm systems should clearly distinguish between alarm, acknowledged alarm, and no alarm (normal conditions).

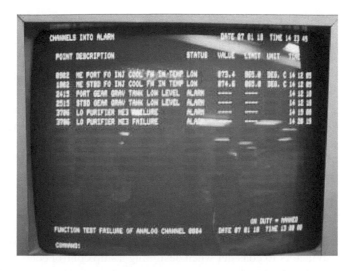

FIGURE 9.22   Alarm list shown on a computer screen. (Personal photograph of M. Lundh.)

- Alarms should be maintained until they are acknowledged.
- Alarms and acknowledged alarms should only be capable of being cancelled if the alarm condition is rectified. This cancellation should only be possible at the individual equipment.
- The number of alarms should be minimised.
- Alarms should be indicated in order of sequence and provided with aids for decision making. An explanation or justification of an alarm should be available (on request).
- The presentation of alarms should be clear, distinctive, unambiguous, and consistent.
- All required alarms should be presented through both visual and auditory means.

9.6.2   VISUAL ALARMS

- Visual alarms should clearly differ from routine information on displays.
- Visual alarms should be flashing. The flashing display should change to steady display upon acknowledgement.
- Acknowledged alarms should be presented by steady display.
- Alarm indicators should be designed to show no light in normal conditions (no alarm) or should be nonexistent on displays.
- Flashing visual alarms should be illuminated for at least 50% of the cycle and have a pulse frequency in the range of 0.5 Hz to 1.5 Hz.
- Visual alarms on the navigating bridge should not interfere with night vision (not applicable in the ECR).

9.6.3 AUDIBLE ALARMS

- Audible alarms should be used simultaneously with visual alarms.
- Audible alarms should cease upon acknowledgement.
- Audible alarms should be differentiated from routine signals, such as bells, buzzers, and normal operation noises.
- Under normal working conditions, the alarm signals should be able to be heard properly inside the wheelhouse and outside on the bridge wings, and their sound characteristics should not be inconvenient to the human ear (not applicable in the ECR). This point is underspecified, and difficult to judge whether it has been fulfilled.

9.7   HEALTH HAZARDS

As an orientation, a list of health hazards is included from the ATOMOS project. These are also factors to consider when designing a ship control centre. Any design project must deal with identification, assessment, and amelioration of short- or long-term hazards to health occurring as a result of normal operation of the SCC. The following hazards should be considered:

- *Noise/vibration*—Continuous/impulse sound or vibration that causes damage to hearing or vibration injuries in the short or long term. The values and references that cover the conditions that spaces on board ship should meet are to be found in the IMO Code for 'Noise levels on board ship'.
- *Toxicity*—Poisonous materials or fumes generated by equipment, capable of causing injury or death in the short or long term. Also consider allergies.
- *Electrical*—Equipment that may provide easy exposure to electrical shock.
- *Mechanical*—Exposed equipment with moving parts that are capable of causing injury.
- *Musculoskeletal*—Tasks that adversely affect either the muscles or skeleton separately or in combination, e.g., lifting of heavy weights, repetitive movements, incorrect disposition of displays and/or commands, etc.
- *Heat/cold*—Sources that provide potential hazards from equipment generation.
- *Optical*—Equipment that is likely to provide ocular injury.
- *Electromagnetic radiation*—Other electromagnetic sources; e.g., magnetic fields, microwaves, etc.

Many of these considerations are studied under several MARPOL (The International Convention for the Prevention of Pollutions from Ships) chapters, where information needed to correctly handle dangerous goods is given (ATOMOS, 1998).

9.8   SYSTEM SAFETY

A key issue for system safety is to identify and understand the factors that affect human performance in relation to the technical systems being operated and the environment in which work is taking place. This task should start from the early stage of the SCC definition and should refine its results as the design progresses, giving the necessary retrofits at different levels: from changes in concept definition to requirement modifications/extensions. Human error analysis should also be integrated with the traditional engineering approach during the phases of the overall safety lifecycle. The activities under this domain should, at least, consider the following key aspects (ATOMOS, 1998):

- *Error sources*—The use of the SCC in general, or of one of its subsystems, which is likely to lead to error; for example, long, complex procedures for simple operations.
- *Use behaviour*—Misuse and abuse of subsystems that have safety implications for the user. For example, inadequate materials, skill and attitude of the system's operator, ergonomic design, and the interpretation of information received are all aspects that have a direct influence on checking human error.
- *Surroundings*—External environmental conditions that have safety implications for the SCC user or third parties involved in ship's operations, e.g., piracy, extreme weather, dangerous cargo (chemical, biological, explosive, and flammable).

9.8.1 SURVIVABILITY

Personnel survivability refers to using system design features that improve safety and operational success while in hostile natural or man-made environments. This includes the progression from the integrity of crew and passenger compartments, through safety, survival, escape and rescue systems, equipment, and procedures (ATOMOS, 1998).

9.8.2 JOINT COMMUNICATION

To sum up this chapter, we point out that we have been discussing two units, the bridge and the engine control room, with the same goals—safety for the crew, ship, and cargo. Other important goals were discussed earlier in this chapter. For these reasons, we must take a systems view on ship safety and the design of ship control centres. Necessary steps along the way are: continued work to generate more guidance, especially for the ECR design, and providing assistance for adapting existing regulations and guidelines to individual ships. This is especially true for human factors and ergonomics guidelines, as the knowledge about how to design good workplaces does exist today, but not always in a form available and applicable to those designing ships. Just as the bridge and the engine room personnel need to communicate in order to do a good job, so do naval architects, researchers, and domain experts, including those with maritime experience. Put together a good mix of competencies in a project group, make sure they understand each other, and a good ship will follow.

ABBREVIATIONS

ATOMOS: Advanced Technology to Optimise Maritime Operational Safety
BRM: Bridge Resource Management
ERM: Engine Room Resource Management
FMECA: Failure Mode, Effects, and Criticality Analysis
IACS: International Association of Classification Societies
SCC: Ship Control Centre
SIC: Station In Control; relates to transferring control between work stations or SCCs
SOLAS: International Convention for Safety of Life at Sea
IMO: International Maritime Organisation
ISO: International Organisation for Standardisation
MSC: Maritime Safety Committee
MARPOL: The International Convention for the Prevention of Pollutions from Ships

ACKNOWLEDGMENTS

A note of thanks to Brian Sherwood-Jones and Erik Styhr Petersen for their assistance. Thank you to all of the cadets in the MTO-Sea Project for the photographs and information.

REFERENCES AND FURTHER READING

Andersson, M., and Lützhöft, M. (2007). *Engine Control Rooms—Human Factors.* Paper presented at the RINA Human Factors in Ship Design, Safety & Operation IV, London, 21–22 March.

ATOMOS. (1998). Conceptual Standard for SCC Design (including HMI) A217.00.11.052.001E. Available at http://www.control.auc.dk/atomos/.

Blackwell, H.R. (1959). Specification of Interior Illumination Levels. *Journal of Illumination Engineering* 54, 317–53.

Blackwell, H.R., and Blackwell, O.M. (1968). The Effect of Illumination Quantity upon the Performance of Different Visual Tasks. *Journal of Illumination Engineering* 64, 143–52.

Blackwell, H.R., Fry, G.A., and Pritchard, B.S. (1963). Reflected Glare. *Journal of Illumination Engineering* 58, 120–27.

Blomberg, O., Lützhöft, M., and Nyce, J.M. (2005). *AIS and the Loss of Public Information.* Paper presented at the COMPIT: Fourth International Conference on Computer and IT Applications in the Maritime Industries, Hamburg, 8–11 May.

Clarke, A. (1978). Coping with the Human Factor. *Marine Design International, Supplement to Marine Week.* March 31, 1978, 23–24.

Courteney, H.Y. (1996). *Practising What We Preach.* Paper presented at the Proceedings of the First International Conference on Engineering Psychology and Cognitive Ergonomics, Stratford-upon-Avon, U.K.

Hederström, H., and Gyldén, S. (1992). *Safer Navigation in the '90s—Integrated Bridge Systems.* Paper presented at the SASMEX, Safety at Sea and Marine Electronics Conference, London, 7–9 April.

IACS. (2007). IACS Recommendation No. 95 for Application of SOLAS V/15. www.iacs.org. uk: International Association of Classification Societies.

IMO. (1974). *SOLAS: International Convention for the Safety of Life at Sea*. London: International Maritime Organisation.

IMO. (2000). *MSC/Circ.982 Guidelines on Ergonomic Criteria for Bridge Equipment and Layout*. International Maritime Organisation.

ISO. (2007). *8648:2007 (E) Ships and Marine Technology: Ship's Bridge Layout and Associated Equipment, Requirements and Guidelines*. Switzerland: International Organisation for Standardisation.

Istance, H. (1978). *Design of a Manoeuvre Module and Its Evaluation by Means of a Static Mock-up*, Report 5316:3. Gothenburg, Sweden: Swedish Ship Research Foundation.

Istance, H., and Ivergård, T. (1978). Ergonomics and Reliability in the Ship Handling System. *SSF Report 157, Project 5311*. Göteborg, Sweden: Stiftelsen Svensk Skeppsforskning (SSF).

Ivergård, T. (1976). Bridge Design and Reliability: An Ergonomic Questionnaire Study. *SSF Project 5311:13*. Göteborg, Sweden: Stiftelsen Svensk Skeppsforskning (SSF).

Lützhöft, M. (2004). *'The Technology Is Great When It Works': Maritime Technology and Human Integration on the Ship's Bridge*. Unpublished Ph.D. thesis, Linköping University. Available at: http://www.divaportal.org.

Lützhöft, M., Sherwood-Jones, B., Earthy, J.V., and Bergquist, C. (2007). *MTO-Sea: Competent Cadets Make Safer Systems*. Paper presented at the RINA Human Factors in Ship Design, Safety and Operation IV, London, 21–22 March.

Mayfield, T.F., and Clarke, A.A. (1977). *The Ship's Bridge and Wheelhouse Ergonomics Design Study*. Paper presented at the HF in the Design and Operation of Ships, Gothenburg, Sweden.

Millar, I.C., and Clarke, A.A. (1978). *Recent Developments in the Design of Ships' Bridges*. Paper presented at the Proceedings of the Symposium on the Design of Ships' Bridges, London, 30 November.

Proceedings of the Institute of Navigation National Maritime Meeting. (1977). The Maritime Institute of Technology and Graduate Studies. Linthicum Heights, MD: The Institute of Navigation, Washington, DC.

Proceedings of the Symposium on the Design of Ships' Bridges. (1978). London: Royal Institution of Naval Architects, Nautical Institute.

Sherwood-Jones, B., Earthy, J.V., Fort, E., and Gould, D. (2006). *Improving the Design and Management of Alarm Systems*. Paper presented at the World Maritime Technology Conference, London, 6–10 March.

Wagner, E. (1994). *System Interface Design—A Broader Perspective*. Lund, Sweden: Studentlitteratur.

Walraven, P.L., and Lazet, A. (1964). Human Factors in Bridge and Chartroom Design. *Journal of Navigation* 17(4), 405–7.

Wilkinson, G.R. (1971). *Wheelhouse and Bridge Design—A Shipbuilder's Appraisal*, Vol. 113. London: Transactions of the Royal Institution of Naval Architects.

Part V

The Human Dimension in the Control Room

10 The Operator's Abilities and Limitations

Toni Ivergård and Brian Hunt

CONTENTS

10.1 INTRODUCTION

The abilities and limitations of the operator in control room work are discussed in more detail in this chapter. We aim to give concrete data that can be of use in the design of control rooms. Much of the background discussion, motivation, and other important aspects is omitted in order to bring about a fully comprehensive solution, as it is not possible to summarise these in a handbook of this type. For more detailed information, the reader should refer to the specialist psychological or sociopsychological literature, or consult ergonomic or psychology experts in the field (for example, Stevens, 1975; Hamilton, 1983; Salvendy, 1986).

The description of human psychology and physiology is also highly simplified. Models of the type given in this book have a certain justification in the planning of control rooms and the like. However, they must be used with care. The discussion is based on the model shown in Figure 2.17. The operator's sense organs and perception are considered first and then the decision and memory functions are examined. In conjunction with these, we give examples of decision-making speed, different types of decisions, and the problems associated with over- and underloading. Some information on the human motor system and ability to carry out various manoeuvring and control movements is given in the later part of this chapter. Chapters 3, 4, and 5 provide examples on the design of information and control devices.

10.2   SENSE ORGANS AND PERCEPTION

In order to be able to control a machine of any sort, the operator needs some form of 'measuring device' (sense organs) to receive information from the information displays. For humans, these measuring devices perceive signals giving the information necessary to operate and control the process.

Table 10.1 and Table 10.2 summarise the capacity and types of the human sense organs. From these tables it may be seen that there are two characteristics of the human sense organs that are very important. First, most human sense organs span a very large range of amplitudes. The eye, for example, can cover an amplitude range that is 10 times higher than the threshold level for sight (the lowest visible light level). The sense of hearing can hear sounds that are 10 to 12 times the threshold level.

The tables also show the very wide frequency range that can be sensed by the different sense organs. However, despite the great amplitude and frequency ranges of the sense organs, there are still stimuli that lie outside the human detection ability.

TABLE 10.1
Some Stimuli, the Sense Organs, and Their Effects

The stimulus affecting the	Sense of organ that produces	The effect
Electromagnetic radiation of wavelengths 300–1500 nm	Eye	Sight
Pressure variations in the air with frequencies between 20–20 000 Hz	Ear	Hearing
Temperature changes	Cold and warmth receptors	Heat and cold
Movement of the limbs	Muscle spindles	(Probably no conscious feeling)
	Golgi tendon organs	(Probably no conscious feeling)
	Tendon receptors	Limb movement
Many gases	Olfactory organs	Smell
Touch	Touch and pain receptors	Pressure

TABLE 10.2
Functional Range of the Sense Organs

Sense Organ	Intensity	Frequency of Wavelength
Eye	Ca. 3×10^{-10} to 0.3 cd/m^2	300 nm to 1500 nm
Ear	10^{-6} to 10^{-2} j cm^{-2}	20 Hz to 20 000 Hz
Vibration sense	2.5×10^{-7} to 2.5×10^{-4} j	Up to about 10 000 Hz

For the human senses there are certain electromagnetic waves that cannot be seen and certain sounds (mechanical waves) that cannot be heard. There are also certain types and qualities of stimuli (for example, ionising radiation, relative humidity) for which the human sense organs are inadequate.

It is easiest for humans to discriminate a signal if it can be compared with another signal at the same time. This is called relative discrimination. Absolute discrimination, where the size of a signal has to be determined directly without having another one to compare it with, is considerably more difficult.

Apart from recognising the frequency and amplitude of signals, human beings can sense and discriminate other dimensions such as locational and temporal relationships. Examples of this are a pulsing sound signal or a flashing light. Table 10.3 gives examples of the ability to recognise different types/variants of one and the same stimulus, and the ability to discriminate quantitatively different types of stimuli. More detailed recommendations on these stimuli are given in Chapter 3.

10.2.1 SIGHT AND VISION

The most important factors that determine the efficiency of the sense of sight are:

1. *Luminance Discrimination*—The ability to distinguish differences and variations in brightness.
2. *Sharpness*—The ability to define and differentiate shapes.
3. *Temporal Visual Ability*—The ability to distinguish changes and movements over time.
4. *Depth Discrimination*—The ability to judge depth (variation in image between the two eyes).
5. *Colour Discrimination*—The ability to discriminate the wavelength of the light.

The first factor is thought to be the most important. The ability to discriminate brightness is usually expressed as contrast sensitivity. Contrast can be expressed by the following formula:

$$K = (L_B - L_S) / L_B \qquad (10.1)$$

where L_B stands for the brightness of the background (luminance) and Ls for the luminance of the signal or object that is to be distinguished. Figure 10.1 shows how visual performance varies with contrast and background luminance. The visual

TABLE 10.3

Numbers of Recognizable Variations of Different Stimuli

Stimuli	Max.	Recommended	Notes
Colours			
Lamps	10	3	Good
Surfaces	50	7–8	Good
Design			
Letters and numbers	∞	?	
Geometrical	15	5	
Figures/diagrams	30	10	Good
Size			
Surfaces	6	3	Satisfactory
Length	6	3	Satisfactory
Lightness	4	2	Poor
Frequency (flashing)	4	2	Poor
Slope			
(Direction of pointer on dial)	24	12	Good
Sound			
Frequency	(Large)	5	Good
Loudness	(Large)	2	Satisfactory

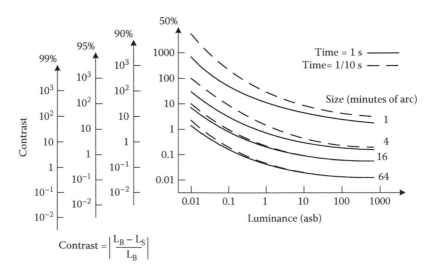

$$\text{Contrast} = \left| \frac{L_B - L_S}{L_B} \right|$$

FIGURE 10.1 The relationship between contrast for different performance levels on the vertical axis (95% is taken to be sufficiently good for most tasks) and luminance on the horizontal axis. The range of curves is shown for different exposure times and sizes of visual target.

efficiency is normally directly related to the lighting intensity and the luminance of the background and surrounding surfaces. It can be seen from Figure 10.1 that for a given contrast (for example, $K = 0.1$), the lighting level must be increased, which also directly increases background luminance, if performance is to be improved. This is because, among other things, increased intensity of light gives increased sharpness of vision.

The requirement of sharpness of vision can be reduced by increasing the size of the critical parts of the visual task. For small objects or dim lighting, recognition speed (temporal visual ability) may be reduced if the light levels are increased. In order to get good visual conditions, it is better to increase the size and contrast of the object rather than increasing the lighting level. If it is necessary to increase the light level, problems of glare must be taken into account. In particular, one must take care to avoid reflections on panels and instrument glasses.

The sensitivity of the eyes changes (adapts) to the average luminance of those surfaces to which they are exposed. The adaptation that the sense of sight carries out means that humans do not experience the luminance of different surfaces in the same way as a light meter. The latter measures only the physical intensity of the luminance. The human eye reacts differently. If, for example, it is adapted for a very low level of luminance and lighting (such as a very large opening of the pupil and very sensitive retina), a surface with a relatively low luminance (but higher than the very low adaptation luminance) is perceived as being very bright. If, on the other hand, the adaptation luminance is very high (for example, due to sitting and looking at a bright sky), even surfaces with relatively high luminances will appear to be dark. These relationships are illustrated in Figure 10.2.

In order to work in the dark, total dark adaptation can be achieved if sufficient time is spent in a dark place. This may occur in certain types of control rooms, for example, on board ships, in certain types of traffic monitoring, and other situations where radar screens are used. In practice, this means that the light sensitivity of the eye increases considerably, and a certain level of visual ability can be maintained despite the darkness. However, it takes up to thirty minutes for dark adaptation to be fully achieved.

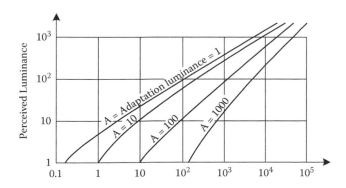

FIGURE 10.2 Relationships between perceived luminance, true luminance, and adaptation luminance. (After Hopkinson and Collins, 1970. With permission.)

It is important that such areas are sufficiently dark and the eye must not be exposed to any direct light. Only red light can be tolerated. The visual acuity (sharpness) is very much reduced when the eye is dark-adapted. This means that only large objects can be seen. Direct reading of small details on instruments in the dark requires the use of special equipment. Certain recommendations on the design of such control rooms are given in Chapter 5.

10.2.2 COLOUR VISION

Colour is not a physical quantity but a psychological one (see Figure 10.3). Electromagnetic waves are converted by the eyes and the visual nerve centre, including the visual parts of the brain cortex, and are experienced by humans in different ways. These different experiences are called colour. In the human retina there are thought to be three different types of sensitive bodies that convert electromagnetic radiation of different wavelengths and brightness, and thereby bring about different experiences of colour.

Table 10.4 gives the approximate relationship between the different physical characteristics of the radiation and the parameters in colour vision. The wavelength of the light is related to the colour experienced (for example, red, blue, yellow, green). The luminance is related to the brightness and the purity of the colour to the perceived colour density. But it is important to remember that there is no simple relationship between the physical characteristics of the light and the mental perception of colour. Colour perception is influenced by a large range of different factors and some of these are examined below.

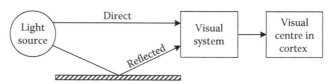

Pigmented surface (absorbed and reflected radiation)

FIGURE 10.3 Relation between light radiation and subjective response to colour.

TABLE 10.4
Approximate Relationship between Physical Characteristics and Colour Perception

Physical Characteristic	Colour Perception
Wavelength	Colour
	blue, yellow, etc.
Luminance	Brightness
'Purity'	Colour density

There are three main sources of colour generation:

1. Light reflected towards and absorbed by a pigmented surface.
2. Electromagnetic radiation (including invisible ultraviolet [UV] light) that is reflected and absorbed by a fluorescent material and thereby generates new light.
3. Direct light, for example, a lamp or a cathode ray tube (TV or visual display unit [VDU] screen).

All these different types of colour generation depend on whether the lighting in the room is coloured. Table 10.5 shows how surfaces with different types of colour pigment are experienced when they are exposed to red, blue, green, and yellow lighting, respectively. A pigment that is experienced as blue (that is, has a 'blue colour' pigment) in 'white' light will be experienced as reddish-purple in yellow light.

A similar but less marked effect is also obtained when fluorescent material is illuminated with coloured light.

As colour is to a large extent a psychological experience that is formed within the human visual system, there are a number of physiological and psychological factors that determine how we perceive colours (see Figure 10.3 and Figure 10.4). The following relatively important factors will be discussed in more detail:

1. Central vision
2. Focusing of colours in the lens
3. Colour blindness
4. Meaning and perception
5. Age

10.2.3 CENTRAL VISION

Figure 10.5 shows a cross-section of a human eye. The incoming light is focused by the lens and falls onto the retina that lines the back of the eye. Different parts of the retina have different types of light-sensitive cells on them. In the central section

TABLE 10.5
Coloured Light on Different Coloured Surfaces

Surface/ Colour	Red	Blue	Green	Yellow
Red	Bright red	Bluish red	Yellow-red	Light red
Blue	Red-purple	Bright blue	Dark green-blue	Reddish purple
Green	Olive green	Green-blue	Bright green	Yellow-green
Yellow	Red-orange	Reddish brown	Greenish yellow	Bright orange
Brown	Brown-red	Blue-brown	Dark olive brown	Brownish orange

Source: Modified from Woodson (1981).

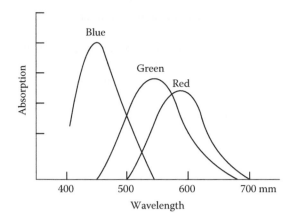

FIGURE 10.4   Relative sensitivity of the eye for different wavelengths of light.

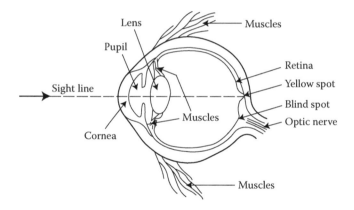

FIGURE 10.5   Cross-section of the human eye.

(fovea) are cones (see Figure 10.6), which are less sensitive to light. The cones are used for detailed imaging, and have a very different ability to distinguish detail from the rods that are found more peripherally around the retina; the rods are considerably more sensitive to light. The rods are used mainly for night vision and the cones for daylight vision. The cones help us distinguish colour.

10.2.4 FOCUSING OF COLOURS ON THE LENS

All types of lenses, including the lens in the human eye, focus light of different wavelengths to different extents; blue is focused nearer the lens and red farther away from the lens. This means that if light containing a large difference in wavelengths is presented to the eye—for example, violet-blue and red on the same VDU screen—the eye will find it difficult to focus at a suitable 'distance setting' from the screen. The colours may be perceived to lie in different planes, and the eye will have to work

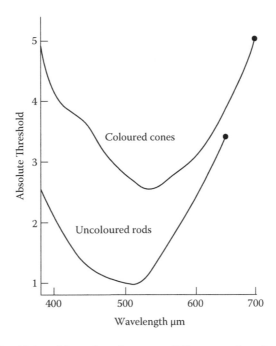

FIGURE 10.6   Sensitivity of the rods and cones to different wavelengths of light.

constantly in order to focus and achieve optimal clarity. One must therefore avoid having colours widely separated in wavelength adjacent to each other.

The perceived colour from a directly radiating light source such as a lamp or a cathode ray tube (TV/VDU screen) will not be affected much by the colour of the surrounding light. It is, however, important to remember that the colour of light perceived on a VDU screen will also be affected by the intensity and colour of the room lighting even if this effect is not very marked. This is because the human eye is affected directly and indirectly by the light in the room, and will become adapted both to its intensity and colour. In addition, there will always be a certain amount of light reflected towards the surface of the VDU screen, and this will be 'mixed' with the light from the screen itself.

10.2.5 COLOUR BLINDNESS

Colour blindness is an important factor that must be taken into account in the design of colour VDU screens. About 8% of all males and 0.4% of all females have some form of colour vision deficiency. Table 10.6 shows the ways in which light seen by someone with normal vision as red, green, blue, or yellow would be seen by people with various forms of colour vision defect. The different descriptions are simplifications, but they do give an indication of the problems that may arise. People with reduced ability of the green receptors (nearly 5% of males and almost 0.4% of females) will experience yellow as green-red, and if the red receptors are inactive,

TABLE 10.6

How Colours Are Perceived by People with Different Colour Vision Defects

Type of Colour Vision Defect[a]	Red	Green	Blue	Yellow
Absence of colour receptors 0.003% men, 0.002% women	White-grey	White-grey	White-grey	White-grey
Inactive green receptors 1.1% men, 0.01% women	Red	White	Blue	Light red
Inactive red receptors 1.0% men, 0.02% women	Whitish	Green	Blue	Light green
Inactive blue receptors 0.03% men, 0.001% women	Red	Green	White	White
Weak green receptors 4.9% men, 0.38% women	Red	Light green	Blue	Green-red
Weak red receptors 1% men, 0.02% women	Light red	Green	Blue	Green-yellow
Weak blue receptors (very rare)	Red	Green	Light blue	Light yellow

[a] *8% of men and 0.4% of women have some deficiency in colour vision.*
Source: Summarised from IBM (1979).

yellow is seen as light green and red as whitish. In the total absence of colour receptors, all light is seen as white, and the world is only experienced on a grey scale.

10.2.6 MEANING AND PERCEPTION

Colours have different meanings for different individuals. Colours are often described as warm or cold, as calming or creating aggression. However, research has shown that there is no uniform way in which people interpret subjectively the characteristics of different colours. The meaning of different colours is constrained by cultural, social, and educational factors. This is also strongly affected by individual personality. It is therefore almost impossible to give any general advice in this respect. There is, however, a certain tendency for the density of a colour to be associated with rising and falling. One could thus use different colour densities of blue to represent different gas pressures, and use different densities of red to indicate temperature levels.

10.2.7 THE EFFECTS OF AGING

Increasing age brings decrements in various aspects of visual ability. For example, speed accommodation (distance setting) is reduced considerably with age. The eyes of older people therefore become sensitive to light of widely differing wavelengths on the same picture, resulting in an accommodation problem. In addition, the ability to discriminate luminance differences and between colours with shorter wavelengths, such as blue and green, is impaired.

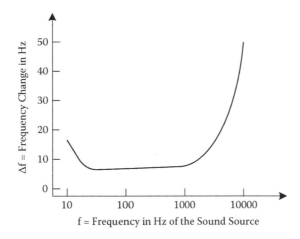

FIGURE 10.7   The ability of hearing to distinguish changes in frequency depending on the frequency of the sound source.

10.2.8 HEARING

The organs of hearing are in certain respects better adapted to receiving signals than are those of sight. This applies particularly to the recognition of complex patterns. Even in the presence of noise and complex sounds from a machine installation, a trained ear can easily detect deviations and diagnose faults. The ability of the ear to differentiate frequencies is shown in Figure 10.7. This shows the smallest change in the frequency of the sound that can be distinguished against the frequency of the original sound. Figure 10.7 shows that it is easier to distinguish changes in frequency at lower frequencies. Changes in loudness, however, can be heard better at higher frequencies.

In order to hear a sound in a noisy* environment, it is clear that the sound must be a louder noise than when there is a quiet environment. Background noise masks the sound signals. Low-frequency noise masking occurs when the masking sound is of the same frequency. The masking effect is increased if the background sound has a lower frequency than the sound to be heard. The masking effect of a noise on speech is expressed in speech interference level (SIL) units. SIL values are based on octave band measurements and are often the same as the current masking noise. The SIL value is the average of the sound pressure levels (in dB) in three octave bands with centre frequencies of 500, 1000, and 2000 Hz. Table 10.7 shows the masking of speech at different distances between speaker and listener.

A hearing aid can improve reduced hearing ability. However, if there are wide variations in noise levels, the hearing aid can actually increase the problem by amplifying all the sound equally.

The need for amplification for those with hearing damage may vary with the intensity and frequency of the sound. Certain people with hearing impairments can

---

* Noise is defined as unwanted sound. Industrial noise and the quietly dripping tap would both be experienced by most people as noise. A sound that may be music to some people may by others be considered as noise.

TABLE 10.7
The Masking Effect of Noise on Speech

	Voice Level in SIL (dB)			
Distance (m)	Normal	Raised	Very Loud	Shouting
0.5	71	77	83	89
1.0	65	71	77	83
2.0	59	65	71	77
3.0	55	61	67	73
4.0	53	59	65	71
5.0	51	57	63	69
6.0	49	55	61	67
7.0	43	45	55	61

hear loud sounds relatively well, while not hearing softer sounds at all. Amplification of the sound can thus mean that the hearing-impaired person would certainly be able to hear the weak sounds, but at the same time the loud sounds would all become too loud. In work situations, therefore, earmuffs are to be preferred. They will attenuate the surrounding noise while the relevant signals will be heard clearly.

10.2.9 OTHER SENSE ORGANS

Human beings have a number of other sense organs, such as the senses of taste, touch, and smell. These, however, are not of any great use as information channels in the control room. Their importance may be for warning—for example, of dangerous gases—but one should never rely on the sense of smell for this. Of the other sense organs, those found in the muscles are the most complex. These, too, are not of any great use as information channels in control room work. However, due to the direct feedback they give, they can be of some importance, for example, in keyboard work or the use of foot pedals in vehicle driving. A driver pressing the accelerator pedal obtains feedback information not only from the gain in speed but also from the receptors in the leg and foot that indicate by how much the pedal has been depressed.

An experienced car driver knows (perceives unconsciously) immediately the meaning of a certain movement on the accelerator. The driver thus gets information directly from the receptors in the foot and leg that indicate the speed of the car a moment later. This means that the driver does not need to wait to see what the speed of the car will actually be. In the same way, the skilled manual craftsman gets information directly from the various limbs without needing to wait to see the consequences of the skilled movements.

10.3   JOB SKILLS

In the modern control room, the operator will generally have some of the following tasks:

1. Start up/stop the system
2. Control, manoeuvre, and regulate
3. Check, monitor (act only when there is a fault)
4. Keep records and report
5. Repair and maintain
6. Plan, programme, and analyse

Different types of skill and knowledge are required in order to do these tasks. Job skills may be divided roughly into seven areas. The first, *perceptive skill*, is concerned with the ability to notice and distinguish signals from complex patterns. These are normally light or sound signals. One example of a difficult perceptual task is radar work. Many inspection tasks, such as the supervision of electronic circuit room work, also set high requirements on perceptual ability. In control room work, hearing is the sense traditionally used to bring small changes in the condition of the process to the operator's attention.

The demands on perception in control room work are sometimes relatively high. They can, however, be reduced if instruments, VDU screens, and light fittings are designed to provide an overall solution. It is important that a good perceptual solution not only enables the various information-giving devices to be easily seen but also that their logical relationships enhance understanding of the process.

The demands on different forms of *cognitive/thinking ability* vary considerably in different types of control rooms. Sometimes the requirements are very small, for example, when the control room tasks are mainly supervisory. When the job involves planning and analysis work, the demand on mental processes may be considerable. It is important for these types of demands to be matched to the operator's abilities, earlier experience, and knowledge. It is also an advantage if the demands are changed in line with the increase in experience over time.

Many modern computerised control rooms have reduced the demands on the memory and its ability. This is because the computer can store any information, tabulations, and coefficients to which the operator may need access. However, under-stimulation can occur because of the special demands of control room work that often involve long periods of low activity. This in turn requires *vigilance ability*, the ability to maintain a certain form of alertness despite the soporific nature of the work. In repair and maintenance work, what is known as *diagnostic skill* is often needed. There are different strategies for looking for faults in technical systems; one can, for example, test different components according to the level of probability that they are faulty. Another strategy is first to test half of the components, and then to carry out tests on the others. Diagnosis skill is often defined as:

1. The ability to find suitable guidelines.
2. The ability to interpret the guidelines in relation to the situation in hand.
3. The ability to select effective testing and searching strategies.

Diagnostic ability is associated with the operator's breadth of job skill. Repairing different faults, even electrical ones, requires not just electrical knowledge but also mechanical knowledge. The requirement for *sensorimotor ability* is considerable in many manual jobs and in vehicle driving. The need for *motor ability* is greatest in craft and purely manual tasks. Control room work does require a certain degree of motor and sensorimotor abilities, but the motor requirements are relatively limited.

In certain types of control room work, where the operator starts and stops the process directly and where the process speed is relatively slow, the demand for *control skill* can be very high. Control skill can be seen as a skill that people possess to varying degrees depending on their experience and other factors. It is thought that the most skilful control operators have some form of 'intuition' of what is the most suitable control movement to make or not make. It is not always thought, however, that it is such people who have the best ability to control and manoeuvre the process. This 'intuition' may be described as the operators having their own conceptual (mental) model of the relationships between control movements and the final result. This conceptual model is a simplified and schematic description of the actual functions in the process (see Chapter 2).

10.4   PERCEPTION AND SIMPLE DECISIONS

The ability of human beings to discriminate signals and the length of time taken to arrive at different forms of decisions are considered in this section. Finally, we discuss the way in which ageing and training can affect decision making.

10.4.1 DISCRIMINATION OF SIGNALS

If there are 10 lamps lit in a room, and one more is lit, the increase is perceived as being considerably greater than if there had been 100 lamps lit in the room and one more had been lit. The same is true for weight; if a 1 kg weight is added to another 1 kg weight, the weight change is felt as considerably greater than if a 1 kg weight is added to 10 kg.

Two researchers, Weber and Fechner (reported in Stevens, 1975), developed a law based on these perceptions. They stated that the difference between stimuli that were just perceptible (JND, just noticeable difference) were always of the same psychological strength. A certain psychological JND should thus cause the same psychological effect as another JND under different conditions. If the JND in weight increase above 1 kg is 1 g and 100 kg is 100 g, the psychological effect of the first increase (1 g) should be as great as the second increase (100 g). In practice, this means that a small increase in a low-intensity stimulus creates the psychological experience as a considerably greater increase in a high-intensity stimulus. The law produced by Weber and Fechner from this background can be expressed thus:

$$n = K \log I \qquad\qquad (10.2)$$

where I is the signal intensity and n is the number of JNDs and K is a constant that depends on the type of signal. In other words, n is a form of index of the psychological perception of the signal.

The psychologist, Stevens, designed another index that is somewhat easier to use. According to Stevens (1975), the relationship between the psychological perception P and the intensity of the signal can be expressed thus:

$$P = LI^a \tag{10.3}$$

L and a are two coefficients that depend on the characteristics of the signal. As Figure 10.8 shows, L represents the point at which the curve meets the vertical axis, and a is the slope of the line.

Stevens' law has been used for describing the psychological perception of a number of different factors such as light signals, sound signals, and lifting of weights. Some examples of the values of the exponent a for various types of tasks are shown in Table 10.8.

10.4.2 Simple Decisions

The time taken to react to a stimulus by carrying out a certain action is called *reaction time* (also called response time). This could be, for example, pressing a button when a lamp lights. The greatest part of the reaction time is taken up centrally in the brain, that is, in making the decision to press the button. The time it takes for the signal to get from the eye to the brain and from the central parts of the brain out to the muscles that are to be contracted constitutes only a small part of the total reaction time. Variations in reaction time vary with different sense organs. For hearing and touch, the reaction time is about 140 ms. For vision it is about 180 ms if the person is prepared for the signal. If someone is not prepared, the reaction time increases considerably. Reaction time for smell and taste are about 500 to 1000 ms. For warning signals where a rapid response is important, it is best to use the senses of hearing or touch, or to use these two senses together.

Reaction time (RT) reduces when the amplitude (strength) of the signal increases (see Figure 10.9a). The degree of motivation also affects RT (Figure 10.9b). If it is important for the operator to respond quickly, RT is reduced. Rapid reactions

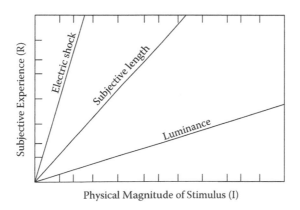

FIGURE 10.8 Stevens' power law on logarithmic scales.

TABLE 10.8

Examples of Values of the Exponent for Various Types of Tasks

Sensation	Exponent	Stimulus
Sound level	0.67	Sound pressure from 3000 Hz tone
Vibration	0.6	Amplitude at 250 Hz on finger
Visual length	1.0	Drawn line
Visual area	0.7	Drawn square
Taste	1.4	Salt
Smell	0.6	Heptane
Cold	1.0	Metal contact on arm
Heat	1.6	Metal contact on arm
Muscle power	1.7	Static contraction
Weight	1.45	Lifting weights
Electrical shock	3.5	Current through finger
Luminance	0.3	Point source
Taste	0.8	Saccharine
Visual speed	1.0	Moving light object

can also be trained. Figure 10.9c shows how the reaction time is reduced for each repeated attempt. Figure 10.9 also shows that the variation between different operators can also be reduced by training. The difference between the best and the worst will thus be less after more training (Figure 10.9c). Reaction time is also dependent on age (Figure 10.9d).

The factor that most affects reaction time, however, is the characteristics of the information to which one has to respond. Shorter reaction times are alarm signals (sound or lamp). If, on the other hand, one has to decide which alarm is active when there are several to choose from, then the reaction time will increase considerably. The same is true if instruments have to be read before a decision can be made. In a simple case, where the choice is between a number of alternatives (for example, choosing between different alarms), the reaction time increases with the logarithm of the number of possible alternatives (Figure 10.9e).

There is also a relationship between reaction time and how important it is to react correctly when choosing one signal from many. If it is important to be correct, the reaction time will increase. Similarly, if one has to respond to several rapid signals consecutively, reaction time increases successively. In order to be able to maintain the short reaction time, the signals must not arrive more often than three short reaction times, three per second, and if the signals are difficult to manage there should be only two signals per second. This is because the situation is rarely simple enough that it is only a question of reading off a signal and making an immediate decision.

In the inspection of sawn timber, several seconds are needed to sort each plank into different groups using a very simple classification. If the signals—the planks—are fed in automatically at too high a speed, the number of errors will increase.

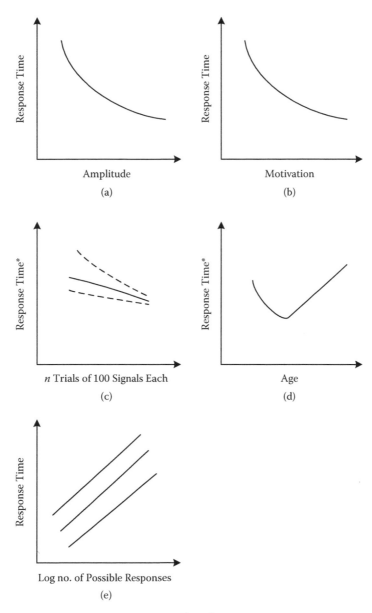

*NB. Response time falls asymptotically in these cases.

FIGURE 10.9 Variation in response time with different factors.

Human beings have a limited decision-making capacity, being able to make only one decision at a time. The human decision-making function is usually described as a one-channel system with a certain limited throughput rate. Signals that follow too quickly after the previous ones have to await their turn in the short-term memory. Decisions are made on groups of information—'parts'. In a control room setting, the size of these groups depends on the operator's skill. This is clearly noticeable in keyboard work. A person with long experience of keyboard work and who uses touch typing does not respond to individual signals (letters) but reads groups of symbols and letters and makes decisions on the basis of these. If the operator notices that something is wrong while typing in, the whole decision is carried out and any corrections made afterwards.

In summary, it may be said that people cannot deal with more than a maximum of one signal per 300 ms, or about three signals per second. Discontinuity is also evident when people attempt to follow a curving line that is continually varying in position. They work with small bits of the curve at a time and produce something similar to a polygon. Somewhat simplified, it can be said the people work in steps, where each step takes about 3/10 of a second. In order to follow a changing curve such as a sine wave, about six steps are needed in order to produce a recognisable likeness. As each step takes 3/10 of a second, it therefore takes 2 seconds per period. In other words, a human being cannot follow variations that have a higher frequency than half a period per second (Hz). As there are 2π radians/cycle, the human bandwidth is limited to between zero and three radians/second. It is important to note the fact that human beings work relatively slowly and lack the ability to follow rapid excursions. This is of great importance when a human element is used as the controller in control systems of different types. Where control events occur faster than three radians/second, the controller cannot cope with them and cannot therefore be used as the controller in such situations.

In order to monitor movements where each movement is a response to a signal, the human limit for movement speed lies at three responses/second. This, however, is not the absolute upper limit for movement; if several movements are programmed with well-trained tasks, the periodic speed can be considerably higher. Typing speeds can be up to about 10 cycles/second. In this case, the typist makes a decision on the movement programme, where each programme contains several movements. Up to three programme decisions can be made per second. In practice, however, decision speeds are considerably lower, but in certain cases at least two decisions per second can be made.

For very rapid movements, the muscles work differently, physiologically speaking, from when slow movements are made. Among other factors, some of the internal feedback mechanisms that exist in muscles and that help to make movements stable and controlled are disconnected in very fast movements. This means that precision and reliability are lower in fast movements than they are in slow movements.

10.4.3 CHANGES WITH AGE

Various aspects of both physical and mental performance decline with age (see Figure 10.10, continuous lines) with maximal ability occurring between 18 and 25 years of age. However, some comprehensive research (Forsman, 1966) has shown that certain types of mental abilities remain fairly stable until a relatively advanced age. In certain cases, some of those types of mental abilities may even show an improvement.

The factors that usually decline with age are known as unstable factors, while those that do not change are called stable factors. Unstable factors include those

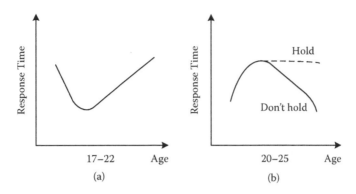

FIGURE 10.10 Changes in performance level with age.

concerned, for example, with short-term memory, such as the immediate repetition of a number of digits. A young person can usually remember six or seven digits without difficulty, and repeat them a short time later; for the same person 10 years later it can be considerably more difficult. The stable factors are those that depend on experience and for which there is no particular time limit for carrying them out.

The sense organs are also affected by increasing age; for example, visual acuity and dark-adaptation ability are reduced, the ability to differentiate tones decreases markedly, and the hearing threshold rises with increased age. Central processes such as reaction time increase and the ability to react to several signals at the same time is very much reduced with age. Physical strength drops and older people are also more sensitive to stress from their surroundings.

Older people have more difficulty translating from one code to another than do younger people. It is, for example, harder for an older person to start driving a left-hand drive car if he or she has always driven a right-hand drive car. It should, however, be emphasised that it is possible to teach new jobs and skills to older people if the teaching methods are suitably designed for their needs. In particular, one must remember that older people need a longer time for training. One must also be careful not to put up social barriers to training; for example, the types of books used in the education of schoolchildren should not be used with adults. Older people also have a greater need to test and prove themselves in their training.

Even though the sense organs and the motor functions change and deteriorate with the years, it is not these input and output mechanisms that are the most important factors for reduced performance in older people. These can very easily be compensated for with good ergonomic design of the environment and control panels and with individual aids such as spectacles and hearing aids. Increasing age does not particularly affect the greater part of attention and motor organisation. For certain types of job skills, it is possible that the abilities may be maintained relatively well because of increased experience.

The most important changes for the older person occur in the central processes. Decision making in different situations takes considerably longer for the more elderly. It is in this area that the greatest limitations occur for the older person in his or her working life, but it is also in this area that ergonomic measures can make it easier.

Age has relatively little effect on more complex decisions where there is little time limitation for carrying out the task, which is common in control room work. Instead, experience and motivation are the most important factors for efficiency. Older people can thus, in practice, often make difficult decisions as well as or better than younger people. There is also much to suggest that the character of the work can aid in the maintenance of a high-effectiveness capability with increased age. An interesting, stimulating, and variable job increases the likelihood of maintaining efficiency. Reduction in productivity also varies much from person to person and these variations are especially great at higher ages (over fifty-five to sixty years).

10.5   OVERLOADING AND UNDERLOADING

In this section we examine some general aspects of fatigue and alertness. Then we discuss problems of underloading, which are especially common in control room work. Finally, we give a short example of the need for variety in the job and the risk of monotony due to lack of variation.

10.5.1 FATIGUE AND ALERTNESS

Fatigue is a difficult concept to define and there are several different types (Astrand and Rodahl, 1977). Purely muscular fatigue is due to the build-up of lactic acid in the muscles that is experienced in more severe cases as pain in the muscles. In addition, the function and efficiency of the muscles is reduced, particularly the ability to perform 'precision' work. The risk of muscular fatigue is limited in modern control rooms, where the degree of muscular loading is usually low. The work can normally be carried out from a sitting posture, or in a combination of sitting and standing. Changes in the degree of alertness are another form of fatigue that occurs due to changes in the central nervous system.

One factor affecting the degree of alertness is the time of day. 'Productivity' or efficiency varies with the state of alertness, although different aspects of efficiency vary with time of day and degree of alertness. Normally, alertness is lowest early in the morning and highest later in the evening. There are, however, large individual variations here too. For some people, the times of lowest alertness is earlier, for others, later. For other people, the variation between the lowest and highest levels of alertness is relatively small. Figure 10.11 shows how efficiency and ability vary with the time of day. Another important factor that affects the state of alertness is the character of the task. This will be discussed in more detail later.

Figure 10.12 shows how performance ability varies with degree of alertness. This shows that there is an optimal degree of alertness, where the performance ability is at its best. If the degree of alertness is low, performance is also low; if, on the other hand, the level of alertness is very high, performance ability may also be low.

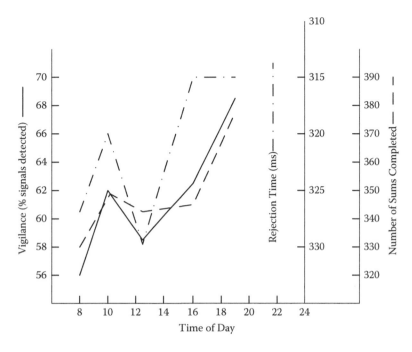

FIGURE 10.11   Efficiency and ability variation with the time of day.

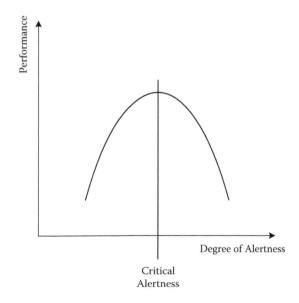

FIGURE 10.12   Relationship between performance ability and alertness.

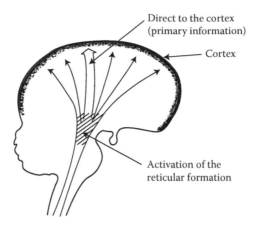

FIGURE 10.13 Diagrammatic cross-section of the brain showing secondary activation of the cortex via the reticular formation in the brain stem. At first, the activation causes an increase in alertness but then disturbs the primary information.

Tasks containing very little variation or jobs with very low levels of loading cause in turn a low level of alertness, and thereby reduced performance ability.

The alertness level can be affected by factors other than those connected with the job. In monotonous work with a low level of stimulation, alertness can be kept at a high level by means of a secondary stimulation, such as music. On the other hand, where the job provides a certain level of stimulation, and thereby a relatively good level of alertness, a secondary stimulation may result in too high a level of alertness and a reduction in efficiency. Music and other secondary stimuli while people work is thus only of value in tasks that are monotonous or where the mental load from the job is low.

Purely physiologically, the degree of alertness is affected by areas in the cortex of the brain. Signals from the various sense organs go first directly to their special areas in the cortex. In addition, signals go via a secondary route to the brain stem and bring about a general stimulation of the whole cortex. This general stimulation prepares the brain to receive signals and information from the various sense organs. If the secondary stimulation is too intense, there is a risk that it will not just stimulate but overstimulate and cause confusion in the interpretation of the primary information (see Figure 10.13).

Control room work is often characterised by low mental loading over long periods, thereby producing a low level of alertness. In addition, this type of work is often done on a shift basis, which means that people are often working during the early morning hours when alertness is naturally and physiologically low. The risk of poor performance is thus very great.

10.5.2 UNDERLOADING

Monitoring work is a typical example of a task causing a low level of loading. The tasks need an operator's level of attention to be kept relatively high, while at the same time there is relatively little for the operator to do and very little is happening. This is usually known as vigilance work. It is typical of vigilance tasks that performance drops significantly after only a short time at the job.

Figure 10.14 shows the performance ability in a typical vigilance working situation. In the alert situation, where there are plenty of signals from instruments and test lamps, the number of signals read and detected correctly may perhaps lie in the region of 99%. If, on the other hand, there are few signals, and these arrive randomly and are weak, the performance will start to drop considerably to 60% to 70% of its original level after as few as 20 to 30 minutes. The characteristic of the vigilance working situation is that the number of signals (such as relevant pointer movements and alarm signals) is very small. In addition, the signals are relatively weak and irregular. A common feature of control room work is the monitoring of signals that occur seldom and irregularly. On the other hand, the strength of these signals may be relatively powerful. Performance may thus be consciously raised by making sure that all important signals are particularly clear-cut. The risks occur when, for example, one has to be attentive to small changes on an instrument pointer. If such small changes are important to the operator, they should be provided with alarms. In this way, the special problems associated with vigilance can be avoided.

In a control room, work loading is commonly very low, often lying below the level required for the job itself to provide a certain degree of stimulation and alertness. Occasionally, however, there are peaks in the loading. These peaks are sudden, often cause load levels well above the suitable level, and are even sometimes above the handling abilities of the operator (see Figure 13.5 in Chapter 13). Peak loadings usually occur in conjunction with some form of disruption in the system. Lower

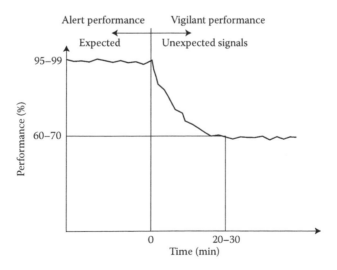

FIGURE 10.14   The ability to perform a vigilance task can be kept high where signals are expected (so-called alert performance) but drops rapidly when signals are not expected (vigilant performance).

levels of increase occur with different types of recording and reporting work, for example, during the start-up of the installation.

It is difficult for people to cope with these sudden peaks when disruption occurs. The difficulty is also greater because the previous loading has been at a relatively low level. These factors produce a lowered level of readiness to handle the loading peaks. If instruments and equipment are unsuitably designed from an ergonomic point of view, or there are different types of environmental loading in the room, such as glare from lighting or disturbing noise, these will make it even more difficult for people to cope with the sudden peak loads. Good environmental design and suitable design of instruments and controls in accordance with ergonomic principles make it easier for the operator to cope with these peaks. It is better to try to reduce the size of the loading peaks so that the operator can organise the tasks into an order of priority during peak load. The working situation should provide clear indications of the jobs to be done first at such times.

Work with a very low level of loading is experienced as being monotonous. But there are also other reasons why jobs can be felt to be monotonous. If the variations in the tasks are low, they will be experienced as monotonous even if the loading level is relatively high. Such is often the case in control room work. In order to raise the loading level in the control room, it is common for the operator to carry out various supplementary tasks, such as reading off and recording from various instruments or filling in various reports and log sheets. However, in themselves, these tasks are rather tedious and are felt to be monotonous. Thus they do relatively little to increase operator alertness. Such tasks should not be expected to aid increased efficiency or job satisfaction.

Supplementary control room tasks given to raise the loading level during low-load periods should be work that provides sufficient variation for the operator. It is important for the extra tasks to be relevant and they should not interfere with other parts of the job. These additional tasks should preferably prepare the operator to cope with the workload peaks. In practice this means that they in some way allow the operator to work on the system and the process. For example, planning work regarding future running of the process may be suitable. Certain maintenance and service work in the process or control system may also be meaningful.

10.6   MOTOR FUNCTIONS

In control room work, demands on motor functions are relatively small. Specialised control rooms such as cockpits on aircraft or bridges on ships have very particular operator requirements. Motor tasks comprise the turning of levers and wheels, filling in forms, pressing keys on a keyboard, and dialling (telephones). As seen from Chapter 5, there are also ergonomic guidelines for how different types of control devices should be designed in order to suit the majority of people. In certain cases the operators need to make fine adjustments. There is often plenty of time in which to do such tasks. One can thus make sure that the control device has sufficient gearing for fine adjustments to be made without too great a motor demand. Some types of work on VDU screens also demand fine adjustments for the positioning of cursors.

In these cases, too, the 'gearing' should also be sufficiently large so that not too great a motor demand is made.

Concrete recommendations on these points are given in Chapter 5. Where rapid movements and fast reactions are required, the muscles work in a different way from slow movements. Keyboards and press-buttons are thus better than wheels and levers where fast control manoeuvres have to be carried out.

Training of motor functions requires mainly practical training in simulators or in real on-the-job situations. Many modern control rooms rely completely on the use of ordinary computer keyboards. In most situations this might be acceptable. However, one important disadvantage with keyboards is the lack of qualitative sensory motor feedback (the only feedback they provide is done/not done). In high workload and critical or emergency situations, this can become very important. Training in rare and critical situations demanding sensory motor actions should preferably be done in simulators and by the use of specially designed controls that are capable of providing sensory motor feedback.

REFERENCES AND FURTHER READING

Astrad, O., and Rodahl, K. (1977). *Textbook of Work Physiology*. New York: McGraw-Hill Books.

Ekman, G. (1959). *Psykologi*. Stockholm: Almqvist & Wiksell.

Fitts, P.M., and Posner, M.I. (1967). *Human Performance*. Belmont: Brooks.

Forsman, S. (1966). Ålder och arbetsanpassning. In M.E. Lutman and H.S. Spencer (eds.), *Handbok i ergonomi*, 569–81. Stockholm: Almqvist and Wiksell.

Galanter, E. (1966). *Textbook of Elementary Psychology*. San Francisco: Holden Day.

Hamilton, V. (1983). *The Cognitive Structures and Processes of Human Motivation and Personality*. London: John Wiley & Son.

Hopkinson, R.G., and Collins, J.B. (1970). *The Ergonomics of Lighting*. London: Macdonalds Technical and Scientific Books.

IBM. (1979). *Human Factors of Workstations with Display Terminals*. G320-6102. New York: IBM.

Kalsbeek, J.W.H., and Ettema, J.H. (1963). Scored Regularity of the Heart Rate Pattern and the Measurement of Perceptual or Mental Load. *Ergonomics* 6, 306–7.

Mackworth, N.H. (1950). *Research in the Measurement of Human Performance*, MRC Report 268. London: HMSO.

Moray, N. (ed.). (1979). *Mental Workload: Its Theory and Measurement*. New York: Plenum Press.

Nickerson, R.S., Eldkind, J.I., and Carbonell, J.R. (1968). Human Factors and the Design of Time-sharing Computer Systems. *Human Factors* 10, 127–34.

Salvendy, G. (1986). *Handbook of Human Factors and Engineering*. New York: John Wiley & Sons.

Smith, Dennis. (2000). On a Wing and a Prayer? Exploring the Human Components of Technological Failure. *Systems Research and Behavioral Science* 17, 6 (November–December), 543–59.

Stanton, N.A., and Ashleigh, M.J. (2000). A Field Study of Team Working in a New Human Supervisory Control Room. *Ergonomics* 43, 8, 1190–1209.

Stanton, Neville, Hedge, Alan, Brookhuis, Karel, Sala, Eduardo, and Hendrick, Hal (eds.). (2004). *Handbook of Human Factors and Methods*. Boca Raton, FL: CRC Press.

Stevens, S.S. (1995). *Psychophysics*. New York: John Wiley & Sons.

Welford, A.T. (1968). *Fundamentals of Skill*. London: Methuen.

Welford, A.T. (1973). Stress and Performance. *Ergonomics* 16, 5, 567–80.
Welford, A.T. (1976). *Skilled Performance*. Dallas, TX: Scott, Foresman.
Woodson, W.E. (1981). *Human Factors Design Handbook*. New York: McGraw-Hill.

11 Learning and Creativity at Work

Toni Ivergård and Brian Hunt

CONTENTS

11.1 SIMILARITIES BETWEEN LEARNING AND CREATIVITY

There is much evidence that learning is facilitated by a free and open environment. This is what is often called a 'high-ceiling' or an error-tolerant environment (Florida, 2006). Many authors and researchers claim this type of environment will also facilitate creativity and innovation (Ekvall, 1996; Florida, 2006). It is therefore understandable that in the minds of many people learning and creativity are one and the same. In a modern way of looking at learning we understand that learning has to be a kind of creative process. The learner needs to reinvent the new knowledge to internalise this in order to obtain the ownership of the knowledge. The iterative intellectual enquiry as an inherent part of a creative process is in many ways similar to our understanding of an efficient process of learning (Schön, 1982; Ivergård, 2000).

11.1.1 PERSPECTIVE

In this chapter we describe the environment of control centres from the perspective of facilitating learning and the creation of new knowledge and skills. We are also look at the perspective of functional and process engineering development. The sources of this learning and creativity are the control centre individuals themselves. Daily experiences in the control centre allow the operatives ideal opportunities to learn and apply that learning towards improving work tasks, the work environment, and system design. Such learning opportunities increase when operatives work in teams and have the possibility of seeing work tasks from several related perspectives.

We describe briefly the development of control rooms, from the add-on work resources of past generations to the high-technology control centres of the current day. Although the physical environments and work conditions may have changed, it is noticeable that, in general, there has been a lag in making improvements in how control rooms are managed. Nowadays, control room operatives and technicians may be educated to higher levels of abilities and have the benefit of belonging to the Internet generation that regards itself as technology-savvy. However, too frequently neglected is the possibility of using these skills to optimise the environment for control room work.

11.1.2 SHIFTING THE PARADIGM: FROM CONTROL ROOMS TO CONTROL CENTRES

Sixty years ago control rooms did not exist. In manufacturing industries workers and supervisors together spent all their work time out on the shop floor. Forty years ago, in the 1960s, separate control rooms didn't really exist in the sense we understand this concept today. In the noisy industrial workplaces of the day, operators and maintenance workers worked for short periods from a purpose-built cabin. These cabins were stand-alone additions to the industrial shop floor. Often constructed of wood or metal, these cabins allowed the maintenance crews to store their equipment. The main purpose of the cabins was to protect the crews from noise and air pollutions of the surrounding work environment.

Constructing separate work cabins for workers with specialised duties satisfied the very important objectives of protection from the noisy environment. This was especially so in the Scandinavian countries. Here, legislation existed to protect the health and safety of industrial workers, building on long-standing traditions of trust and cooperation between workers and employers. In this environment, acceptance of such legislation was ensured as its intent was mutually understood. However, one disadvantage was that by working in the cabins, the workers or operators became distanced from the 'shop floor' and the machinery and processes they were supposed to supervise and maintain. Contemporary research showed that there were differences in behaviour between old and young workers: older workers spent much more time out on the shop floor than their younger workmates (Ivergård, Istance, and Günther, 1980).

Twenty years later, in the 1980s, we began to see special control rooms where operators could supervise, and to some extent control, industrial processes. Early examples were electrical power production and distribution. Inside the control room the process flows were represented in visual displays. When process flows reached a critical stage (for example, when pressures became extreme or a risky situation was imminent), the control panel alerted the operator by emitting visual or auditory warnings such as flashing lights and/or sounds. At about this same time, control centres in the pulp and paper industries also came along on a large scale. In most, if not all, of these uses, the main driving force was to improve process control and reliability and to avoid total breakdown of the process. In industry, restarts are time consuming and very costly. In countries where electrical supply is inconsistent, power outages are obviously extremely costly and disruptive.

11.2   EDUCATION, TRAINING, AND LEARNING AS A PART OF DAILY WORK

A definitive characteristic that differentiates humans from machines is the human ability to adapt to different situations. While some adaptation takes place instantly, such as when a person withdraws his or her hand from a hot surface, some adaptation may take place over a relatively long time. When adaptation is done systematically and according to a particular methodology, the concepts of education and training are used. In an education and training programme for job skills, the first stage is usually the 'needs' analysis during which the requirements of the job in question are determined. Traditional work-study methods may be of some use here regarding the job content and requirements. However, in control room work the traditional work–study methods are of little use as most work takes place inside the heads of the operators and is difficult to observe. In order to determine the coverage of the tasks in the job and its character, interviews must be carried out with the operators involved. Judgements may also be made based on studies of performance and abilities in similar situations. In this respect, it is also helpful to observe operators at their work. Observation converts tacit knowledge (the innate skills of the operators in practice) into codified knowledge (the observer's new knowledge of how work tasks are carried out by the practitioners). When the practitioners are prepared to share their tacit knowledge by explaining their rationales for making certain decisions and taking certain actions, then the tacit knowledge is processed in double-looped learning (see Schön, 1982; Raelin, 2001; Edmondson, et al., 2003).

The analyses of the job and job requirements that form the basis for the determination of the education and training programme are often the same as those that form the basis of the whole of the personnel planning process. These analyses are also involved in the optimal ergonomic planning. In this connection, it is worth emphasising that the job descriptions and the job requirement analyses can form the basis for selecting the technology used and identifying the best interface (between the machine and the people). One must not therefore bring any prejudices or preconceived notions to the design of training. It is counterproductive to consider how a job or work task must be taught in order to be done under certain definite conditions. Instead, using principles of ergonomics, one could change the conditions and thereby place more focus on the job or task—and thus simplify the learning process. For example, it may be necessary to take steps to change a working situation if it is unsuitable to design a training programme for a certain type of job and a certain type of person. Here the fundamental issue to consider is the match of human operator to the machine. Two key issues relate to the design of training programmes. First, it is important to design effective training that is simple in the sense of easy to follow and understand. Secondly, the training programme should be suitable for the individuals involved in the training. The investigation of needs, operator interviews, and (ideally) observation of operators at work are intended towards this outcome.

Sometimes a related problem is solved by laying off the existing personnel and recruiting newer, often more highly-educated people to fit changes taking place in the production technology. This is a strategy driven by the technology. Obviously, it would be more cost effective to adopt the technology to the people, augmenting their perceived potential, if necessary, by additional training. This approach will be particularly cost effective for the organisation. The approach can also benefit the local society, for example, in sparsely populated areas of a country where employment opportunities are scarce and where there is a limited supply of suitably-educated workers. In this type of situation, it may be difficult to recruit locally a sufficient number of highly-educated workers. One possible alternative might be to encourage personnel to migrate from major cities. However, this will only be a stopgap solution. Over time, this type of personnel may wish to relocate back to their region of origin. To rely on the existing workforce and invest in a kind of continuous lifelong learning is likely to prove a much more efficient and reliable solution in the longer term. This investment in localised skills is likely to bring benefits to the individuals, to the area, to the employer investing locally, and to the government in terms of skills development.

11.3   AIMING FOR SUCCESS IN EDUCATION, TRAINING, AND LEARNING

In the design of an education and training programme of learning, the best results are likely to stem from following a number of ground rules:

- In general, people can only absorb a limited amount of information at any one time. For some types of learning, the content and process of training should therefore be divided up in some way into stages, so that not too much information is presented at once. However, there are exceptions to this general rule of thumb. When learners are expected to build up a tacit knowledge model of an industrial or administrative process, then the content can be holistic and highly information-intensive. This point will be discussed later.
- The demands made on speed and accuracy must be selected very carefully for the different training stages. For certain types of work where accuracy is very important, it is better to insist on high accuracy right from the start; speed can come later as skill level increases. In this case, exemplification of 'good practice' and feedback on the learners' efforts to replicate good practice should clearly focus on accuracy. A good example here is learning a foreign language, where beginners need to make themselves understood to interlocutors at a basic level of pronunciation, vocabulary, and grammar as well as learning the basic sounds and shapes of the new language. For other types of work (including speed skills), the reverse may be better, i.e., to require the highest possible speed at the start and allow low accuracy and to let this improve as learning progresses (Clay, 1964).

- Feedback is of critical importance. The trainee must be informed continually during training of his or her performance at different levels of learning. Over time, some trainees can become adept at using feedback to accelerate their learning.
- As far as possible the trainee must be prevented from using incorrect methods. If a trainee makes a mistake at some point, the trainer should ensure that this is used as feedback to show where improvements can be made. It is not overly helpful if errors are repeated as these may act as negative reinforcement and adversely influence the learning. Only in exceptional cases should the incorrect method ever be demonstrated (and then clearly indicated as 'how not to do it'), as this can easily become entrenched in the learner's memory and mixed up with the correct method.
- The trainee should be able to control the teaching pace in order to match it to his or her own learning progress. This is often overlooked, but is a key feature of so-called learner-centred approaches. The trainees' learning is often positively influenced when the learner has greater control over the pace of learning (see Paulsson, Ivergård, and Hunt, 2005).
- Motivation is also essential for a successful process of learning.

For the learning of mainly manual motor skills (in this context including feet, legs, and other limbs), the ordinary five sense organs do not play an especially large role. More important are other feedback systems such as the natural sensors included in the human muscles that are able to sense changes in tension, speed, and acceleration (a kind of tacit experience melded with tacit knowledge). Thus from the very beginning, it may often be suitable to insist on accuracy being as high as possible. Allow a slow speed to start with, to facilitate the learners' confidence in their performance. As training progresses the speed can be increased (again bearing in mind a learner-centred approach in which learners have control over the pace of learning). In sensory-motor work (for example, a combination of touch and muscular receptors), there are complex control problems that play an active role in coordination of hand movements and muscle activity. In such situations it is often better to maintain a predetermined speed from the start and to let accuracy improve as skill improves. In such a 'lockstep' approach, development of the two skills in parallel will be mutually supportive of the learning.

In mainly intellectual training, it is especially important to be able to control one's own learning pace. Part of the benefit of learners controlling their own pace is to allow reflection on accumulated knowledge (see, for example, Kolb, 1984). Naturally, people have their own preferred styles of learning and it is helpful if trainers can identify these at an early stage. This also assumes that the trainer knows how to facilitate different styles of learning and how best to accommodate these in, for example, syllabus and curriculum design, teaching style, and in practical learning assignments.

There is relatively little information on training and the process of learning for control room operators. However, research has shown that performance of certain types of control room tasks increases if the operator has some technical and factual knowledge of the functioning of the process (Ivergård, Istance, and Günther, 1980).

Another description of training and learning in control room work emphasises the importance of not dividing up the work into too much detail. In one type of process it may be self-evident to the trainee operators what one does when starting up a certain process unit. In such a case it may only be necessary to say 'Start up process X' in the specification. Most people will know how to attend to turning knobs and operating keyboards. It is therefore unnecessary to divide these subtasks into even smaller units such as open the cover, turn the knob, slacken the nut; one need only say 'Change printer page' instead, if this is the job in question. Figure 11.1 shows an example of suitable division of jobs at different levels in order to give a basis for training.

Regarding training programmes for older people, it is important to remember some simple rules:

1. Allow plenty of time for training.
2. Allow self-determination of learning pace.
3. Combine verbal and written information; verbal instructions alone are often inadequate.
4. Provide opportunities for practical work early in the training process (learning by doing).
5. Avoid mistakes occurring early in the training (correcting mistakes/errors by doing it right).

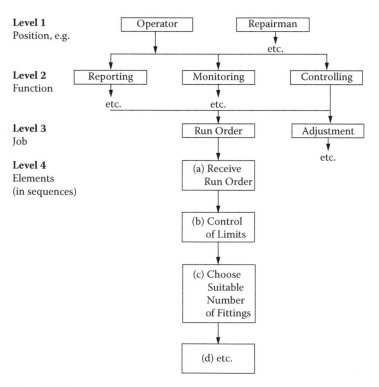

FIGURE 11.1   Division of jobs at different levels to facilitate training.

11.3.1 Simulator Training

An operator can be influenced and changed in different ways in order to improve the performance of the human-machine system. One method is to use various types of selection processes in order to try to identify and select those people who are best suited for the particular type of work. The question of selection will not be dealt with here, primarily because in most cases it is a difficult method to apply, and one whose value is highly disputable. If selection tests are to be used, these need to be properly evaluated with known validity and reliability. However, tolerance to stress might be a personality factor to be considered. More difficult is the issue of introverted and extroverted personalities in control room work. Only relevant case-valid systems should be used. On the other hand, a refined training method is often of great value. Many of the guidelines found in the common educational literature are also applicable to the training of control centre operators. In this context, therefore, the emphasis here will be on the form of training that is of particular interest to the training of operators in process industries.

Traditionally, it has been possible to identify three types of learning: factual learning, attitude modification, and job skill training. To these can be added: skills for lifelong learning, training for team working, skills for self-directed learning, learning how to learn, and creative skills development. It is also important to consider one's own preferred style of learning and being creative. Over the past decades different forms of technology have been developed to support processes of learning. The past decade has seen an explosive development of e-learning and other forms of Web-supported and self-directed learning. This has partly emerged from distance learning using radios, TV, and ordinary postal services. Historically this is also related to simple early technological aids for learning, for example, the introduction of computers created computer-based training (CBT). The more sophisticated techniques were also developed earlier in the form of simulations of processes. Quite soon the nascent aviation industry made use of the early simulation technologies. Aircraft pilots used the first types of simulator training, which were forms of shadow techniques and were used to prepare the pilots for night-time flying. In this technique the learners learned from 'controlling' their own vessel or aircraft in relation to the shadow images of other craft or buildings displayed in front of them. Similar techniques were later also used for the shipping industry and simulating the manoeuvring of ships in difficult approaches, and navigating vessels in narrow archipelagos. In the era of computerisation there have emerged extremely sophisticated simulators for aviation and shipping training. Today there is a very rapid development in the area of using simulators for training and learning.

Use of simulators in aviation training is probably one of the most important areas of applications. For commercial aviation, simulator training has taken over a very large proportion of the training of commercial pilots. In theory they can be used for more or less 100% of the training input. However, in practice, this is not feasible. Traditionally the strength of aviation simulators has been in dynamic skills training. Nowadays simulators are also used for factual learning, for example, in navigation training for both maritime and aviation applications. More recently, simulator training has begun to be used increasingly in more diverse application areas. It has become more common, for example, in maritime training, but its use has also been increasing within other industries, for example, nuclear power facilities, power plants, and electrical distribution centres. Simulator training has a high potential to be used in most types of modern control centres. Simulators can also be used for redesign and development of industrial and administrative processes. Other examples include simulations that replicate market movements in an economy and simulators on which trainees may hone their skills in trading currencies, stocks and shares, and other financial instruments. The potential in this area is very large, particularly if the simulator training or related development tools integrate the use of advanced stochastic mathematics combined with modern software for the creation of simulation programmes.

However, the main use and experience of simulators is in aviation training and development. Here, scientific evaluation has been carried out in order to determine the advantages and disadvantages of using simulators. In other areas, similar scientific research is found only to a more limited extent. A large part of the experience from within aviation can be transferable to other areas with a certain degree of confidence. Experience within simulator training shows that the advantage of simulators is in the training of sensory-motor job skills and the ability to handle different types of procedures (read off instrument A; if the value exceeds X adjust 4, and so on). On the other hand, simulator training has been regarded as less suitable for the more classic types of factual learning or for attitude modification. Nowadays, we know differently (see below).

A prime advantage of simulator training is economic. In simulator training, learning periods can be considerably shortened. This is an attractive feature in industries, such as aviation, where time is money—a factor that is increasingly so in all industries. Simulator training is also important in that it enables training in unusual or dangerous situations that are seldom encountered in real life. By their nature, simulators allow training in especially important situations that could, for instance, lead to serious accidents in real life. Obviously, there is very high potential for simulation training of extreme and very rare situations in economics, in trading, and in energy distribution. Simulator training can be used both for the initial learning and training, and for follow-up training, especially in connection with the unusual real-life situations.

It is, however, important to remember that simulator training can replicate elements of the real world but it can never wholly replace training on the real system. In spite of sophisticated technical advances, a simulator will only be a replication of reality. It will never be the same as the reality itself. A simulator for training in particularly dangerous tasks must resemble reality as closely as possible. In other words very good mathematical models need to be incorporated into the programme design. On the other hand, if it is to be used for training job procedures and routines, the simulator can in most cases be considerably simpler. It may sometimes even be sufficient just to have a simple mock-up without electrical connections (that is, without any realistic responses on instruments to different control movements by the operator). This simple mock-up type of simulator can also be used for factual training, such as memorising start-up routines, names and design of instruments and controls, or emergency procedures.

In the early 1980s, Scandinavian Airlines used simulators for team training and management learning. In this situation, the simulator training proved to be a very good method for attitude changes. In this context, it became possible to educate cockpit crews to accept role demarcation, for example, between the co-pilot and the captain. This usage of simulators for team training has also spread to other areas of learning how to work in groups. In control rooms and other control centres, this is a very important organisational skill in which employees need to be trained. In new types of control centres—for example, in energy trading or other financial centres—this type of training can enhance efficiency and output of the whole organisation.

When planning simulators for training, it is important to take account of educational expertise from simulator training. There is always a risk that in such training one may build in small deviations from reality, which could lead to a 'negative transfer' effect in the training. This may result in someone handling the real system as though it were the simulator, which in turn could result in the operator making critical errors. This risk of negative transfer seems to be larger in very high-tech applications where the simulation very closely resembles real-world experience (that is, the simulation is so good that the user doesn't expect any deviations from real-world conditions).

The training officer who designs the training programme on the simulator must therefore carry out a very careful skills analysis of the real-life job at the beginning, in order to determine which factors in the job are critical from the point of view of education or training. Of particular importance is to be conscious of the long-term need for new skills and also as a process of lifelong learning. Only then is it possible to determine how the education and training programme should be put together. Furthermore, a simulator designed for learning can easily also be used for creative work and for the development of the work process and of the simulator process itself. In the beginning, we discussed the relationship between creativity and the process of learning. Simulator training can to a large extent be combined with creative work. To make this form of combined learning and creativity meaningful, it is necessary to create a high-ceiling environment that at the same time facilitates teamwork.

11.3.2 The Concept of Control Centres

Control centres are ideal places in which to encourage individuals and work teams to develop their own work space. Control rooms usually comprise a tight community of like-minded individuals who usually have similar work backgrounds and experience. This work environment is most likely separate from other work areas and usually allowed to be self-managing. The configuration of the work teams themselves is very important. This makes it possible (even desirable) for the teams to create the knowledge that the team itself needs to function optimally in the work. Impediments to this knowledge creation are likely to originate from outside this close working environment—for example, from the existing practices of the particular workplace, the extant thinking paradigms of the workers themselves, or a mixture of both. Here, we describe an environment of 'high ceilings' where managers practice a high tolerance for ideas that at first might appear outlandish.

In many ways, we use 'high ceiling' as a metaphor for extending the figurative space made available to employees to think and conceive new approaches to their work. Creative thinking (otherwise known as 'out-of-the-box' thinking), is stifled in environments where employees are expected to follow existing behaviours and where managers 'keep the lid on' the work environment. For creativity to thrive, it is necessary to break through the barrier of traditional ways of thinking and doing. Obviously, in most control room situations one cannot compromise on safety issues. This is clearly a reason why the use of simulators is so important, as it is possible in a simulated environment to train and prepare for risky situations.

Control centres using advanced computer systems based on advanced mathematical modelling have only just begun to be developed incorporating the human factor. A key word in this context is financial *engineering* as taught at schools of advanced engineering and mathematics. This blending of analytical finance, stochastic mathematics, and mathematical modelling with process engineering is an excellent example of what we can expect of many types of control centres in the very near future, not only in the domains of finance and trading but in most control centre applications. Modern techniques of software development will make it possible to design and redesign simulation models. In this way simulation models can be used for development purposes at the same time as they are used for training and also in daily operations. When simulation models are used for training they can initiate creative thinking and innovations that will lead to advanced system development.

The energy sector is likely to create a visionary future. There are enormous potentials, opportunities, and ramifications in such diverse areas as energy conservation and environmental protection as well as in the optimisation of business functions and processes. This new trend will give rise to increased concerns about corporate social responsibility (CSR). Most likely it will be important to clearly separate business-related systems from the operational systems. The operational systems can be very large and cover nations or a group of nations. The Scandinavian countries have a highly coordinated system for electrical power production and distribution. At the same time there are large energy business systems operating over the whole of Europe. In the local market there are many other business operators serving the end consumers. For any organisation, the introduction of new technologies is a linchpin for both technological change and for organisational-wide process development and change (see Hope and Hope, 1997). This is particularly so when technologies develop and evolve with such speed as to aid and abet an organisation's activities and encourage a paradigm shift.

11.4   SOME THEORETICAL PERSPECTIVES OF LEARNING AND CREATIVITY AT WORK

We regard action learning as an opportunity to stimulate reflective learning from critical inquiry into the organisation. Four components—conceptualisation, experimentation, experience, and reflection—accord with standard models of learning (for example, Schön, 1982; Kolb, 1984; Raelin, 1997). One such model (Raelin, 1997) is shown in Figure 11.2.

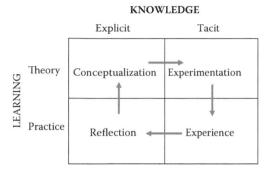

FIGURE 11.2   A model of the learning process. (Raelin, 1997, p. 565. With permission.)

The Raelin (1997) model shows two axes: knowledge and learning. Knowledge is explicit or tacit. Tacit knowledge is the knowledge (know-how, but not know-why) possessed by an individual. Such knowledge is often difficult to explain, to codify, or to set down in media that is accessible by others. Conversely, explicit knowledge can be captured (codified) in databases, instruction manuals, and handbooks. Codified knowledge is available to more people than the people who had the original knowledge. The other dimension of this model is learning, which has theoretical and practical components.

Learning needs to balance between theoretical knowledge (how processes are ideally executed) and practical knowledge (how processes are in fact executed by the people who carry them out as part of their job). The resulting 2x2 matrix has four cells. *Conceptualisation* is explicit knowledge processed through theory. *Experimentation* is the application of theory into tacit knowledge. *Experience* is the practical application of tacit knowledge. *Reflection* is the practical application of explicit knowledge. Progressing through the four-stage learning process (as show by the arrows in Figure 11.2) completes the learning cycle. Peer groups are important to learning. Peer groups provide inputs through evaluation of the work of other people, and offer advice, criticism, and support. In a work group, the inputs from peers aid discussion and encourage mutual learning through constructive criticism, trial and error, and discovery of how others work. A key feature of work groups is the possibility of members learning from each other. In the real world of 'practice' and 'experience', 'tacit knowledge' overlaps with practical experience of 'explicit knowledge'. Practice learning of explicit knowledge is not only from processes of 'reflection'; in reality, reflection is a part of an iterative process with its focus on learning in the area of explicit practice. As presented, the model is cyclical and systematic. This is helpful as an exemplar of the stages in the learning process. However, to facilitate creativity it is helpful to take a more *ad hoc* approach to these stages, for example, by backwards, forwards, and random 'leap-frogging' mental processing of the four different parts of the matrix.

In the work environment of the control room, this model of learning helps to encourage and develop employee-generated learning. Through the efforts of their employees, organisations can develop by learning about learning. Traditionally, organisations that encourage and facilitate employees to learn from themselves and each other have a means of self-generating new knowledge. The knowledge and skills of employees themselves provide the engine for generating new ways of working and thinking about work. Peer-group learning also has a great potential to facilitate creative and innovative work by the operators, who have valid insights into their workplace. If the organisational and workplace environments are supportive, a mutual learning process could encourage creativity.

However, this potential is only effective if employees are allowed to propose solutions to work-based problems. Through their everyday work roles and tasks, employees have valid insights into their workplace. Organisations have opportunities to establish processes of learning that encourage the development of these insights. One example is via processes of collaborative inquiry with others. A key factor in this process is collaborative reflection so that shared learning and shared knowledge generates new knowledge. Collaboration and reflection can transform insights into strategies for action. Workplace teams are able to create and share knowledge among the team's members. These encompass the individual members of the team, the team as a collective unit, and other colleagues who may be members of other similar teams. This in itself might also limit their ability for creative out-of-the-box thinking. Traditional ways of thinking and paradigms will easily limit the scope of thinking. It is necessary to break this barrier. The process of action research must include an approach that allows ways of proposing new ideas and new thinking. The phrase 'high ceilings' is used to describe environments that have a high tolerance for ideas and concepts that at first might look outlandish. This also requires mutual trust among co-workers. It also demands that leaders and managers adopt a more consultative role. To be avoided is the 'I know best' mentality, as this precludes constructive dialogue.

11.5   THE ENVIRONMENT OF MODERN CONTROL CENTRES FOR CREATIVITY AND LEARNING

The environment of modern control centres is well-suited for creativity and learning. For most of the time, the available computer power has an enormous surplus capacity available for other types of application. By default, computer capacity contains surplus dimensions to cover peak loads, for example, system breakdowns or major errors. Apart from being held 'in reserve' to cover such occurrences, this overcapacity can be used for simulations of different types of process applications or possible scenarios of breakdown and errors. For example, simulations could be used for operator training in which the operators practice for future possible scenarios. Simulations could relate to updates of new equipment and systems, including computer systems for process control. However, the control centre operators should also be able to use the available computer capacity for development of the existing production system and also the computer control system, including the related peripherals. Management should emphasise creating a good and motivating high-ceiling environment to facilitate high creativity in this type of development work. Figure 11.3 shows the Raelin model with our addition of the high-ceiling tolerance to encourage learning and creativity.

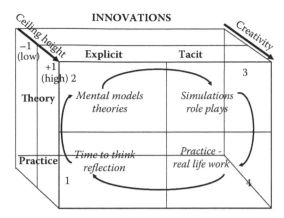

FIGURE 11.3 Adding tolerance as the third dimension for learning. (Developed from Raelin, 1997, p. 565. With permission.)

Added to the model are two possible ceiling heights. In an environment of low tolerance (designated in Figure 11.3 as −1), organisational environments are such that employees are not encouraged to think outside the box (and often, not even to think for themselves). In these types of environments employees become automatons who follow set patterns of work based on extant practices and top–down instructions. New knowledge emanates from the top down, and often remains the preserve of senior managers. Employees are routinely starved of the knowledge they need to fulfil their job tasks. Conversely, in an environment of high tolerance (designated in Figure 11.3 as +1), organisational environments are such that employees are given freedom of thought and action. These types of environments are characterised by employees proposing new solutions to workplace problems. New knowledge is generated from the bottom–up and shared between people who can benefit from that knowledge. New knowledge, in the form of solutions to business problems, is often acted upon by the employees themselves.

11.6   THE CHAIN OF CAUSALITY FOR LEARNING AND CREATIVITY

We now discuss the chain of causality that leads from a high ceiling and tolerance for error to learning and creativity. Figure 11.4 shows the chain of causality.

The chain of causality begins with an organisational environment of high ceilings in which trial and error, experimentation, and the making of mistakes and errors are not only tolerated but are an accepted way in which employees conduct themselves. In such an environment employees are afforded freedoms to make improvements to the ways in which they work. Trial and error and experimentation are two of the ways in which employees discover improvements to their job tasks. Toleration of errors and mistakes are the ways in which the organisation encourages further experimentation. Experimentation in an environment in which mistakes are tolerated leads to two pathways. The first of these, explicit knowledge, is knowledge

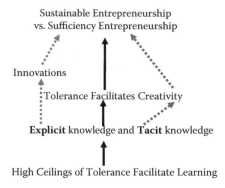

FIGURE 11.4   The chain of causality for learning and creativity.

available to all employees because it is captured explicitly in documents, computer programmes, handbooks, process diagrams, and other guides to workplace practice. Because this knowledge is explicit and available for all employees to use, it leads to innovations within the company. The second pathway, tacit knowledge, is knowledge available to individual employees in their personal skills, competences, and intellect. Thus, this form of knowledge is unavailable to all employees because it is in the minds, hands, and hearts of individuals. Although this form of knowledge is personal rather than public, it nevertheless can lead to creativity. And, as we have described earlier, individual creativity can be shared in work groups to create shared knowledge that is more widespread.

The organisation has three possible pathways to sustainable entrepreneurship. The solid arrows indicate the conventional pathway to sustainable entrepreneurship. One set of dotted arrows leads to sustainable entrepreneurship via innovations generated through explicit knowledge of the collected workforce. The other set of dotted arrows leads to sustainable entrepreneurship via creativity generated through the tacit knowledge of individuals in the workforce.

11.7   ARTIFICIAL INTELLIGENCE, SIMULATION, AND CREATIVITY

Control centres themselves will have built-in advanced artificial intelligence (AI) that will support sustainable entrepreneurship. This will demand a completely new understanding of control room work and its need to be transformed from an environment for supervisory tasks to one where creativity and learning take place in an open environment.

Our working definition of AI is: the ability of machines and other devices to perform activities normally associated with humans, including the ability to modify behaviours on the basis of learning from errors and experience. In the long term, AI could support the operators to build up a database and related systems for process optimisation. It can also provide decision aids for error handling. In this way opportunity will be available for the operator's own creativity and development work that might be outside the scope of AI. It is important to alert the operators to the risk of becoming too dependent on the AI system, because this system will always (or at least mainly) act within the current paradigm (that is, thinking within the box). The big advantage of human operators is their potential to be proactive, creative, and out-of-the-box thinkers. However, a basic condition is that the organisation and its top management and managers on all other levels are willing to accept the challenge of creative thinking. There is always a risk that creative thinking will be perceived as odd, unrealistic, and perhaps (in this context) dangerous. The use of simulations will be an obvious alternative to allow creative thinking and experimentation without intervening in the real system. In other words we are here focusing on the top right-hand box in Figure 11.3. Simulation is a very good example of experimenting with new ideas and theories with the use of tacit knowledge (in this instance, a kind of feeling of new possibilities, which from time to time come into the heads of creative, experienced operators).

11.8   LEARNING FROM THE SCANDINAVIAN EXPERIENCE

In international ratings, Sweden is one of the top countries in the world in the area of creativity, innovation, and knowledge. A report by the World Bank ranked four Scandinavian countries (Sweden, Denmark, and Finland) in the top for knowledge creation (World Bank, 2008). In this same report, Norway and Iceland, other Scandinavian countries, ranked fifth and twelfth. Table 11.1 shows the ranking of the top 12 countries in terms of knowledge creation and innovation.

Richard Florida and Irene Tinagli ranked three Scandinavian countries (Sweden, Denmark, and Finland) in the top five places for Euro-Creativity (Florida and Tinagli, 2004). The Euro-Creativity Index is based on three indicators: Technology, Talent, and Tolerance. An amalgam of these factors, together with other measures, indicates trends in creative capacity.

The Scandinavian countries have relatively more multinational corporations (MNCs) than most other countries in the world. Sweden is frequently at the top. While these types of ratings invariably have weaknesses, they nevertheless provide an overall indication.

11.8.1   SOME SCANDINAVIAN LEADERS IN CREATIVITY AND INNOVATION

A large number of Scandinavian MNCs founded their business operations and products on the basis of some unique innovations. Founded in 1876, Ericsson has become a world leader in communications technology with a business philosophy that communication is a basic human need. With a history of over 130 years of technological innovation, Ericsson has over 20,000 world patents and files over 500 patents per year (see http://www.ericsson.com). In the field of innovative service management and business innovations, Scandinavian Airlines System (SAS) is well renowned. From its first transatlantic flight in August 1946, the airline introduced innovations, including the introduction of tourist class and transpolar routings (see http://www.

TABLE 11.1

Top 12 Country Rankings for Knowledge, Creation, and Innovation (World Bank Data)

Rank	Knowledge Creation	Rank	Innovation
1	Sweden	1	Switzerland
2	Denmark	2	Sweden
3	Finland	3	Finland
4	Netherlands	4	Denmark
5	Norway	5	Singapore
6	Australia	6	Netherlands
7	USA	7	USA
8	Canada	8	Canada
9	Germany	9	Israel
10	Switzerland	10	Taiwan
11	New Zealand	11	UK
12	Iceland	12	Japan

Source: Table constructed from World Bank KAM data for 2008 (http://info.worldbank.org/etools/kam2/KAM).

flysas.com). Later (in the early 1980s) Jan Carlzon introduced the Businessman's Airline and in 1989 SAS University (a place where people learn and develop as a part of their daily work). Founded in 1950 by Ruben Rausing, Tetrapak is a world-class innovator and producer of liquid packaging with technologies and processes that blend hygiene, convenience, and cost saving. The company's motto, 'Packaging should save more than it costs', reveals the company's vision to minimise adverse impacts on the environment (see http://www.tetrapak.com).

From its first showroom in 1953, Ikea (Ingwar Kamprad Elmtaryd Agunnarad) rapidly became a recognised global brand of low-cost furniture that is well designed and functional. Ikea's innovative manufacturing and supply processes ensure that its products satisfy consumer needs (see http://www.ikea.com). Nokia has built well-deserved success on its reputation for consumer design blended with technological innovation. From its beginnings, Nokia has focused on technology leadership and consumer-driven design (see http://www.nokia.com). Each of these companies has developed its position of global leadership over a relatively short time span. In the past decade or so, sustainable large companies continue to be created.

11.8.2 Underlying Rationales for Learning at Work

A key mission of learning at work is to expose control centre operators to an open, fun, and stimulating environment of learning and creativity. While this is intended to be an enjoyable environment, it is also very demanding on both the individuals and the group. The content and structure of any work initiative relating to learning at work is to ensure that operators will be inspired, encouraged, and feel empowered to explore the passion of discovering their own potentials. The learning environment will be a reflection of an environment to excel in creativity and innovation.

The main means of a learning environment is through action learning in project teams. The operators will need to be exposed to different methods and techniques to stimulate creativity and innovation. In this way, they will get insights into how to exploit new ideas and innovations and turn these into successful business opportunities. Professor Emeritus Göran Ekvall from Lund University in Sweden has carried out decades of research on creativity and organisational climate (see, for example, Ekvall, 1997, 2000). In a journal article, Ekvall (1997) identified ten dimensions of organisational climate that influence organisational creativity:

- Challenge
- Freedom
- Idea time
- Dynamics
- Ideas support
- Trust and openness
- Playfulness and humour
- Conflicts
- Debates
- Risk-taking

Ivergård and Hunt evaluated an organisation based on these factors and concluded that organisations with high productivity also have a better organisational climate (see Hunt and Ivergård, 2007).

11.9   SUMMARY AND CONCLUSIONS

In our introductory overview we discussed how nations need to be creative as a way of ensuring investment in their future. Individuals too need to feel they have opportunities to develop and grow. After all, the collective activities of individuals form the backbone that supports the development of a good society. The creativity and innovations of individuals provide a foundation for national development. Companies, governments, and international nongovernmental organisations (NGOs) play important roles in establishing high-ceiling environments in which creativity and innovation can flourish. In high-ceiling environments people feel able to realise their dreams and achieve their potential.

In organisations that emphasise creativity and innovation, the prevailing work environment is an important feature of the workspace. Sensitive managements aim to create an environment where routine practices encourage individuals to develop new ways of thinking and acting. A critical task for management is therefore to establish such an environment and to encourage workers to engage in creativity and innovation. This is best done through example and a commitment to 'walking the talk', that is, putting into practice the message from management. To encourage such an environment, managers should allow employees freedoms of thought and action. This includes setting expectations that do not insist on existing paradigms. Outside-the-box thinking can then become an accepted part of workplace behaviour. In such high-ceiling environments, errors are tolerated as it is recognised that errors are a positive result of experimentation. When their employees make errors, managers in high-ceiling workplaces find ways to reward rather than punish. The new generation of control centres envisages the creation of such an environment as a way of encouraging participants to change their worldview and free their creative passions.

REFERENCES AND FURTHER READING

Baumard, P. (1999). *Tacit Knowledge in Organizations*. Thousand Oaks, CA: Sage.

Checkland, P., and Holwell, S. (1998). Action Research: Its Nature and Validity. *Systemic Practice and Action Research* 11, 1 (February), 9–21.

Clarke, Thomas, and Clegg, Stewart. (1998). *Changing Paradigms: The Transformation of Management Knowledge for the 21st Century*. London: HarperCollins.

Clay, H.M. 1964. *How Research Can Help Training*. London: HMSO.

Edmondson, A., Winslow, A. Bohmer, R., and Pisano, G. (2003). Learning How and Learning What: Effects of Tacit and Codified Knowledge on Performance Improvement Following Technology Adoption. *Decision Sciences* 34, 2 (Spring), 197–223.

Ekvall, Göran. (1996). Organizational Climate for Creativity and Innovation. *European Journal of Creativity and Innovation* 5, 1, 105–23. Also available at http://www.m/creativity.com/map2003/innovat/climate.html.

Ekvall, Göran. (1997). Organizational Conditions and Levels of Creativity. *Creativity and Innovation Management* 6, 4 (December), 195–205.

Ekvall, Göran. (2000). Managerial and Organizational Philosophies and Practices as Stimulants or Blocks to Creative Behaviour: A Study of Engineers. *Creativity and Innovation Management* 9, 2 (June), 94–99.

Ellstrom, Per-Erik. (2001). Integrating Learning and Work: Problems and Prospects. *Human Resource Development Quarterly* 12, 4, 421–36.

Florida, Richard. (2002). *The Rise of the Creative Class: And How It's Transforming Work, Leisure, and Everyday Life*. New York: Basic Books.

Florida, Richard. (2004). America's Looming Creativity Crisis. *Harvard Business Review* 82, 10, 122–36.

Florida, Richard. (2004). *Europe in the Creative Age*. Report published by the Carnegie Mellon Industry Center.

Florida, Richard. (2005). *The Flight of the Creative Class: The New Global Competition for Talent*. New York: Harper Collins.

Hartley, J., and Benington, J. (2000). Co-research: A New Methodology for New Times. *European Journal of Work and Organizational Psychology* 9, 4, 463–76.

Hope, J., and Hope, T. (1997). *Competing in the Third Wave: The Ten Key Management Issues of the Information Age*. Boston, MA: Harvard Business School Press.

Hunt, Brian, and Ivergård, Toni. (2007). Organizational Climate and Workplace Efficiency. *Public Management Review*, 9, 1, 27–47.

Ivergård, Toni. (2000). An Ergonomics Approach for Work in the Next Millennium in an IT World. *Behaviour & Information Technology* 19, 2 (May–June).

Ivergård, T., Istance, H., and Günther, C. (1980). *Datorisering inom Process Industrin, Part 1*. Stockholm: ERGOLAB.

Kasl, E., Marsick, V., and Dechant, K. (1997). Teams as Learners: A Research-based Model of Team Learning. *Journal of Applied Behavioral Science* 33, 2 (June), 227–46.

Kolb, D.A. (1984). *Experimental Learning: Experience as the Source of Learning and Development*. Englewood Cliffs, NJ: Prentice-Hall.

Kristensen, Per, Gustafsson, Anders, and Archer, Trevor. (2004). Harnessing the Creative Potential among Users. *Journal of Product Innovation Management* 21, 1, 4–14.

Kuhn, Thomas. (1970). *The Structure of Scientific Revolutions*. Chicago: University of Chicago Press.

Leonard-Barton, Dorothy. (1995). *Wellsprings of Knowledge*. Boston, MA: Harvard Business School Press.

McKenney, J. (1995). *Waves of Change: Business Evolution through Information Technology*. Boston, MA: Harvard Business School Press.

Montreuil, Sylvie, and Bellemare, Marie. (2001). Ergonomics, Training and Workplace Change: Introduction. *Relations Industrielles* 56, 3 (Summer), 465–69.

Nonaka, I. (1999). The Dynamics of Knowledge Creation. In D. Ruggles and D. Holtshouse (eds.), *The Knowledge Advantage*. Capstone US Business Books.

Nonaka, I., Umemoto, K., and Sasaki, K. (1998). Three Tales of Knowledge-Creating Companies. In G. Von Krogh, J. Roos, and D. Kleine, (eds.), *Knowing in Firms*. London: Sage.

Paulsson, Katarina, Ivergård, Toni, and Hunt, Brian. (2005). Learning at Work: Competence Development or Competence Stress. *Applied Ergonomics* 36, 135–44.

Pisano, G. (1994). Knowledge, Integration, and the Locus of Learning: An Empirical Analysis of Process Development. *Strategic Management Journal* 15 (Winter), 85–100.

Polanyi, M. (1966). *The Tacit Dimension* (reprinted 1983 by Doubleday & Co., New York); partly reproduced in L. Prusak (1977) *Knowledge in Organizations*, Boston: Butterworth-Heinemann, 135–46.

Raelin, J.A. (1997). A Model of Work-Based Learning. *Organizational Science* 8, 6 (December), 563–78.

Raelin, J.A. (2001). Public Reflection as the Basis of Learning. *Management Learning* 32, 1 (March), 11–30.

Reason, Peter (ed.). (1994). *Participation in Human Inquiry*. London: Sage.

Revans, R. (1983). Action Learning: Its Terms and Character. *Management Decision* 21, 1, 39–51.

Sandberg, Karl, Ivergård, Toni, and Vinberg, Stig. (2004). e-Service to Citizens in Rural Areas. *International Journal of the Computer, the Internet and Management* 12, 2 (May–August), 213–22.

Schön, Donald. (1982). *The Reflective Practitioner: How Professionals Think in Action*. New York: Basic Books, Inc.

Singleton, W.T. (1967). Systems Prototype and Design Problems. *Ergonomics* 10, 120–28.

Takeuchi, H., and Nonaka, I. (2004). *Hitotsubashi on Knowledge Management*. Singapore: John Wiley and Sons (Asia).

Tapscott, D., and Caston, A. (1993). *Paradigm Shift: The New Promise of Information Technology*. New York: McGraw-Hill.

Udas, K. (1998). Participatory Action Research as Critical Pedagogy. *Systemic Practice and Action Research* 11, 6, 599–628.

Zuber-Skerrit, O. (2002). The Concept of Action Learning. *Learning Organization* 9, 3/4, 114–24.

12 Modelling and Simulation in the Pulp and Paper Industry
Current State and Future Perspectives

Angeles Blanco, Erik Dahlquist, Johannes Kappen,
Jussi Manninen, Carlos Negro, and Risto Ritala

CONTENTS

12.1 INTRODUCTION

Although the pulp and paper industries have used balancing calculations and process control for a very long time, the scope of modelling and simulation applications is not yet as complete as it could be. One reason is that we seldom have steady-state conditions and we lack adequate measurements for on-line control for many quality variables. Also the final quality of paper depends on many process elements in a complicated and nonlinear way. Control actions for these processes are usually based on skill and experience of the operators. There is a need to improve paper machine performance to increase the competitiveness of the mills. In turn, this makes necessary the application of operator decision-support tools based on dynamic optimisation in order to allow process operators and engineers to manage complex dynamic problems in an efficient and ecologically sustainable manner.

New techniques, such as multivariate statistics, soft-sensors, that is, regression algorithms made from process signals and lab measurements of quality variables, and general stochastic distribution modelling and control in combination with physical models will be used to enhance the controllability of the process. By using and combining these techniques, it will be possible to make further improvements in the papermaking processes. Some of these techniques have already been explored in the pulp and paper industry for data analysis tools, improved process efficiency, development of soft sensors, and stochastic distribution-control algorithms. However, the complexity of the papermaking process makes it difficult to achieve robust models that can be used without frequent work with updating. Producing robust models is therefore a key issue for researchers and suppliers of control systems.

As this is a key issue for the industry, the COST Action E36 'Modelling and Simulation in the Pulp and Paper Industry' (COST, 2005) was established in order to facilitate the coordination of activities and the exchange of knowledge at the European level. The main objective of the COST Action is to advance the development and application of simulation techniques in the pulp and paper manufacturing processes in order to minimise the environmental impact and to increase mill productivity and cost-competitiveness. Among the benefits will be a better understanding of the operation of the processes and their involved control loops. At the current time, the paper industry has a number of pending issues such as increasing stability, improving product performance, achieving higher paper quality, optimising wet end chemistry, better runnability, and lowering environmental impact.

At this moment, 50 researchers from more than 20 organisations from 12 countries participate in the Action. There is also close cooperation with industrial and academic research partners. The following aspects have been considered: available modelling and simulations tools, current use of these tools off-line and on-line, recommendations on the exchange of know-how contained in models, research needs, and main requirements for further software development. There is also a book entitled *The Use of Modelling and Simulation in the Pulp and Paper Industry*, (Dahlquist, 2008).

12.2   MODELLING AND SIMULATION OF PULP AND PAPER PRODUCTION

Pulp and paper production processes are complex and cover most of the unit operations known in chemical engineering. Due to the large amount of water and the low efficiency of most of the separation steps, many recycling streams are required in the process. This is true for water, fibres and fillers, air, and energy. Due to this, the effort to produce a simple steady-state balance can be very high. Over the past two decades, heat balances have been handled with varying success, as have contaminant balances such as, for example, chemical oxygen demand (COD) or macro-stickies. More difficult problems—for example, the simulation of the effects of various contaminants, such as the deposition of pitch and stickies—have not yet been handled successfully within the industry. Simulation of pulping processes has further advanced and can contribute some experience in handling multicomponent balances including complex chemical reactions. Still, an improved steady-state modelling is the key to a better design of pulping and paper production processes.

The paper production process is highly dynamic and although many academic papers have been published in relation to dynamic simulation and paper machines, only a few approaches to the dynamics of the paper machine at an industrial scale have been published. Some studies focus on the dynamics of white water (Orccotoma et al., 1997) and on the dynamics of the wet end at the paper machine (Hauge, Slora, and Lie, 2005). Wet end can be considered as one of the most complex combinations of hydrodynamics and colloidal chemistry. No simulation model has yet been able to fully describe the processes taking place in the wet end. Process simulation is an optimisation tool and only if the dynamics is understood can the dynamic optimisation of the paper production process be addressed (Laperriere and Wasik, 2002).

The activities within this scientific area cover all topics concerning the modelling and simulation of the entire paper production process. This includes the modifications of fibre properties in the stock preparation, the modelling of the complex wet end chemistry, and the modelling of water loops, including chemical reactions and energy balances. Special attention is given to use of dynamic process simulation, real-time simulation tools, and model validation tools.

12.2.1 Availability of Data as an Important Success Factor

The availability of data is an important success factor for any modelling and simulation project. Nowadays mills have access to extensive process data. While there have been high levels of investment to install and operate databases, data management software, and process monitoring systems, the value currently derived from these systems tends to be low. This puts pressure on the promoters of the investment and their common belief that this high amount of data will be a value in itself. In many cases this proves to be wrong. Typical reasons besides a lack of expertise to properly analyse data are that parts of the process are not available to the database system or that the process is not equipped with some perceived need to solve a specific technological question. The 'old-fashioned' sampling technique is still the best practise, especially when dealing with complex wet end problems and when setting up mill balances for pulp, water, energy, and detrimental substances (Hutter and Kappen, 2004)

12.2.2 Methods for the Analysis of Data

Process monitoring systems facilitate the study of historical data and establish trends. However, the possibilities of advanced data treatment for process control and optimisation are not always well known or exploited by papermakers. The first step is to classify the data and the second step is to analyse them in order to answer a specific question. Depending on the final aim, different analysis techniques can be applied, methods being quite often complementary:

1. Classic statistical analysis techniques, time series analysis, experimental design, D-Optimal design, and trial-and-error methods are often used for quick and adequate solution of isolated problems (Box, Hunter, and Hunter, 1978).
2. Building physicochemical deterministic models is the optimum approach when the overall behaviour of the process or one of its sections must be modelled, in order to, for instance, design new equipment or study the kinetics of flocculation (Blanco et al., 2002; Negro et al., 2005; Thomas, Judd, and Fawcett, 1999).
3. The third approach is a combination of approaches (1) and (2) above. Sometimes the physics and/or chemistry of the phenomenon are not known, but a general model has to be built to predict process evolution, optimise process variables, or simulate scenarios. Advanced data analysis tools—e.g., multivariate analysis techniques and artificial neural networks—can provide models with better performance than techniques from approach 1, if the behaviour of the process, section, or equipment to model is complex and unknown (Miyanishi and Shimada, 1998; Masmoudi, 1999; Parrilla and Castellanos, 2003, Blanco et al., 2005).

12.2.3 MODELLING AND SIMULATION SOFTWARE

For modelling and simulation purposes there are numerous software packages currently in use in the pulp and paper industry. In general terms, it is possible to subdivide them into three groups:

- *System or process level modelling*—Object-oriented software is used to conduct flowsheet-based mass balance calculations. Balances and functionalities are often based on first principles and include water, pulp, energy, etc. This is the most commonly-used approach to process description and simulation. In most cases, the approach to process optimisation is still empirical. Many software packages are available. Klemola and Turunen (2001) have published a complete report on Finnish modelling and simulation software. A recent COST report also gives an overview of the current use of software by COST E36 members (Alonso, Blanco, and Negro, 2004). The unit operations are defined as objects in libraries that can be positioned on a worksheet. The relationships between the objects are defined by lines drawn between the objects. The functionality is defined within dialogues of the individual objects. The equation system built in the background is typically calculated with a sequential or a matrix solver (Mathworks, 2002; Goedsche and Bienert, 2002).
- *Unit operation level*—High fidelity models of single pieces of equipment or smaller functional groups. Typical examples are paper machine dryer models. These pieces of software are typically written in plain code like FORTRAN or C++. They either originate from the scientific community or from the equipment suppliers. A new trend is that suppliers try to incorporate these models into the flow-sheet-based software. A breakthrough was achieved by Andritz linking its submodels as .dll files to the process model

and thus providing the full functionality and at the same time safeguarding the undisclosed code (Mayböck, 2004). In line with this, big equipment suppliers like Andritz and Metso have bought companies that develop flow-sheet-oriented software. Now, only a few software suppliers are independently active on the market.

• *Software for data analysis and data-based modelling and simulation*— This type of software is typically applied in cases where no physical models are available or even the question of cause and effect is unresolved. Data analysis followed by modelling can be performed as described earlier. Typical examples are wet end chemistry modelling issues. A typical example for this type of software is Matlab (mathworks) (Mathworks, 2002, 2005).

12.2.4 OFF-LINE USE OF SIMULATION AND SIMULATION-BASED OPTIMISATION

Perhaps the most typical applications of off-line simulation in pulp and paper production processes are various design problems starting from defining elementary mass and energy balances for a process concept with steady-state simulation to using dynamic simulation to integrated design of process and control (Kokko, 2002). As the design task progresses, the model fidelity increases but the scope of simulation often decreases from plant-wide models to subprocesses or individual unit operations. Unit operation development is often supplemented by model development. In this application, models are not necessarily used for getting quantitative results but rather for increasing knowledge on the underlying phenomena behind the unit operation. Examples include use of computational fluid dynamics (CFD) for head-box design (Hämäläinen and Tarvainen, 2002) and wet-press models (Gustafsson and Kaul, 2001; Honkalampi and Kataja, 2002) for press section.

A second important application is operator training using simulators with all the process functions alone or normally also including the complete distributed control system (DCS). In this way the operators can train both to understand the processes as well as the interaction between the processes and the control system (see Chapter 4). The third application is to use modelling and simulation for production support and optimisation. Techniques and challenges for these applications are covered in detail in the next section on on-line simulation. Typical off-line applications include periodic tracking of material balances and process operation improvement, like optimisation of grade changes (Lappalainen et al., 2003). Steady-state tools can be used to keep track of material balances, but for most applications dynamic simulation is the tool of choice. One recent application is maximising total profitability with dynamic production planning (Vänni and Launonen, 2005).

As design and operational problems are often multiobjective by nature, multiobjective optimisation techniques have been lately applied to pulp and paper problems (Hakanen et al., 2004). These techniques allow optimising simultaneously against several criteria and finding balanced solutions without having to constrain some of the criteria *a priori*. One of the most important design and operation criteria is product quality, and models linking the process conditions to the end-product property are still very much under development. Statistical models have been built to estimate the effect of process conditions on the product properties, but due to their nature they are suited for optimisation of an existing pulp or paper production line where enough data can be collected for the models (Blanco et al., 2005; Scott and Shirt, 2001).

One of the complicating facts in defining mechanistic quality models is the complex chemistry of pulp and papermaking. Cooking and bleaching chemistry has been traditionally modelled with kinetic expressions (Andersson, Wilson, and Germgard, 2003), whereas multiphase equilibrium chemistry has been applied successfully lately for scale formation and metal management in pulp manufacturing (Bryant, Samuelsson, and Basta, 2003; Räsänen, 2003; Sundquist et al., 2004) as well as for pH control in neutral paper production (Ylen, 2001).

One of the recent joint efforts in the process modelling and simulation community is to define standard interfaces for the various simulation software components, like solvers, unit operation models, and thermal packages (Computer-Aided Process Engineering [CAPE-OPEN] Laboratories Network, CO-LaN, 2005)). This development makes it, from the user's point of view, possible to mix and match software components from various sources without having to purchase a complete set of software packages. From the vendor's point of view, it obviously gives the possibility to specialise in, for example, model development for the different platforms. Some major players in process simulation, like Aspen Plus, are already partially CAPE-OPEN compliant but to our knowledge no compliance yet exists in the pulp and paper software.

In addition to already referenced papers, some additional off-line applications of simulation models are: maximising total profitability with dynamic production planning by combining production planning and scheduling tools with cost management (Vänni and Launonen, 2005); optimal reutilisation of waste heat in paper mills (Kappen, Mueller, and Kamml, 2004); identification and removal of stickies (Kappen, Hamann, and Cordier, 2004), and so on.

It has to be noted that the above-mentioned examples have been solved using different types of models and tools, and so far no single general tool or set of models exists that can be used for every type of problem. Instead of attempting to build a general tool that would be nonoptimal for most applications, information and model transfer between the tools should be improved to avoid tedious redefining of data. Often process models are built for a specific purpose only—for example, process retrofit—and not utilised later. The biggest advantage of models would be obtained if they were utilised through the whole life-cycle of the plant. The plant model would be then built in the design phase of the project and subsequently used to:

- Check the dimensioning of equipment and the feasibility of control loops.
- Train operators on a system that is based on the simulation model.
- Verify DCS logic; perform DCS checkout based on the model connected to the DCS instead of the real process.

The benefit of this approach is a rather steep start-up curve providing the payback of an investment of US$500,000 to $3 million (Bogo, 2004). Further use could be made in using the model to check on-line sensors and identify the deviation from the set point of individual equipment or the process as a whole. This would call for the model run in real time in parallel to the process itself. A model that covers the whole life-cycle of the plant would require a well-defined and uniform data structure behind it to incorporate both design and operational data and to communicate with process design tools, automation systems, data repositories, and so forth.

12.2.5 ONLINE USE OF SIMULATION AND SIMULATION-BASED OPTIMISATION

Due to the complexity of the pulping process and the dynamics of the paper production process, process control systems have widely been established in the pulp and paper industry. For a typical application the number of the input/output (I/O) connections can vary between 30,000 and more than 100,000. In most cases conventional control technology is used. Operators see the actual values displayed on process displays, proportional, integrational, differential (PID)-controllers help to operate the plant.

Various approaches can be followed to manage these complex systems in an optimal way. The first issue to be addressed is how to handle the huge amount of data available within the system. Fast data acquisition, high-dimensional data analysis, and dimensional reduction play a major role in providing the right data set. In addition, new sensors that have become available within the past few years have to be evaluated as additional sources of information.

Based on the available and digested information, two approaches are to be followed depending on the issue addressed. The first approach is an open-loop decision making supported with simulation. Nonlinear system modelling combined with multivariable system optimisation is one of the basic principles to be used here. Models are calculated based upon calibration with real-data scenarios. The results are available to the operators and the engineers who will then decide about the next steps to be followed, as shown in Figure 12.1.

The second approach is closed-loop control. The performance of all quick control functions—currently carried out with simple PID controllers—is to be

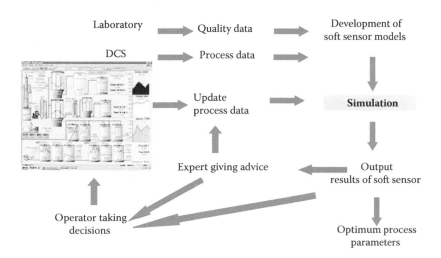

FIGURE 12.1 Modelling and simulation as decision-making support tools.

evaluated. Then, it can be decided whether there is a potential for improvement with advanced control techniques, such as multivariable process control.

Up to now few on-line simulation applications have been realised in industry. Most of the software tools have been developed in other process industry sectors. The petrochemical industry is the leader in applying simulation on-line. Applications in the paper industry are the multivariate model-based optimisation (multiple-input multiple-output–model predictive control [MIMO–MPC]) of paper machine quality control (basis weight, filler, moisture, and so on) in both machine and cross direction that is currently widely implemented in commercial systems (Metso, ABB, and Honeywell). The optimisation of basis weight, coating weight, and moisture in the cross direction of the sheet were actually one of the first (if not the first) high-dimensional MIMO-MPC even though the model of dynamics was rather simple.

These applications serve as a good starting point for optimisation based on advanced simulation. Such ideas have recently been explored in an EU-funded research project (Ritala et al., 2004). The main goal, therefore, is to identify new applications and to use all know-how concerning the dynamic description of paper-making processes to adapt other available solutions to the paper production process.

Some examples described in the references are:

- *Wet end stability*—Multivariable predictive control to model the interactions in the wet end and in the dry section in order to reduce sheet quality variations by more than 50% and to keep quality on target during long breaks, during production changes, and after broke flow changes. It is achieved by coordinating thick stock, filler, and retention aid flows to maintain a uniform basis weight, ashy content, and white water consistency (Williamson, 2005).
- At an integrated mill in Sweden, multivariate analysis (partial least squares [PLS]) was used to form a model to predict fifteen different paper quality properties. First, 254 sensors were used, but 80% of them turned out to be not reliable due to poor calibration, so only 50 were used. After a further analysis we identified twelve variables as most important, of which five were varied in a systematic way to form good prediction models for the fifteen quality variables. The sensors for the twelve most important variables were calibrated frequently during the experiments. The models were made by measuring online data for a certain volume element of fibres flowing through the plant, and adding off-line quality data for the final product. The model then was used online as a predictor to optimise the production. The problem was that the models started to drift after some weeks and became less good. This is a problem with statistical black box models, at least for many variables (Dahlquist et al., 1999).
- An alternative is then to use a grey box model approach, where the model building starts with a physical model, which is tuned with process data. This type of model was shown to give very reliable values on a 'ring crush test' on liner board for several years without any need for recalibration, and was thus used as part of a closed-loop control at several mills (Dahlquist, Wallin, and Dhak, 2004).

- A physical deterministic digester model was made including chemical reactions in a continuous digester as a function of temperature, chemical concentration, and residence time. Circulation loops and extraction lines were included as well. This was used as part of an open-loop MPC and the set points were implemented by the operators manually. The optimised production increased earnings of US$800,000/year in relation to the results if the 'normal production recipe' had been used for the same wood quality and production rate (Jansson and Dahlquist, 2004).
- Web-breaks diagnosis was carried out using feed forward artificial neural networks (ANNs) and principal component analysis. Variable selection and further modelling resulted in several improvements: first-pass retention and first-pass ash retention improved by 3% and 4%; sludge generation from the effluent clarifier was reduced by 1%; alum usage decreased; and the total annual cost reduction was estimated around €1 million (Miyanishi and Shimada, 1998).
- At a Spanish paper mill, multiple regression techniques and ANNs are being used to predict paper quality. From an initial set of more than 7500 process parameters, only 50 were preselected for modelling. Statistical analysis has allowed reducing the number of model inputs to less than ten. Predictions for different quality parameters have been very accurate (e.g., R^2 for paper formation predictions over 0.74). The next step is to optimise model usefulness and robustness through appropriated validation procedures and reduction of the amount of inputs (Blanco et al., 2005).

Figure 12.2 shows how sampling frequency may affect the perception of the information. If the frequency of sampling is low, then true changes may not be determined and lead to an incorrect conclusion of the process performance. The information we have gathered can be used in a decision support system as a Bayesian Network. An example of this is shown in Figure 12.3 and the process display of the same data is shown in Figure 12.4.

In dynamic optimisation carried out in parallel with the process, the quality of on-line data and the dynamic validity of the models are of utmost importance.

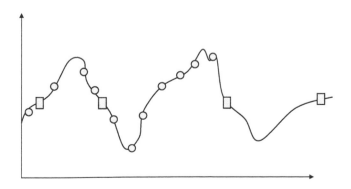

FIGURE 12.2 Sampling frequency and its effects on information.

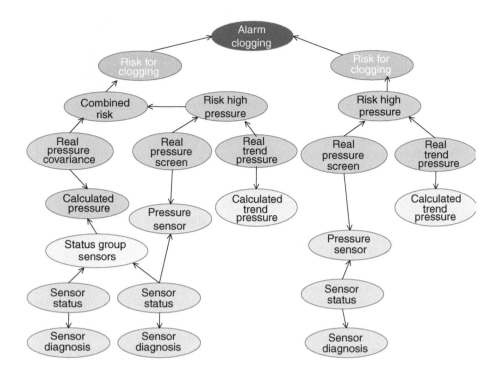

FIGURE 12.3   An example of a Bayesian Network.

Quality of data is severely compromised by the slow changes of the characteristic curve relating the signal (in 4 to 20 mA) to the item measured, such as consistency. The common practise for maintaining measurement quality is that occasionally samples of the processed intermediate or final product are taken to a laboratory to be measured. When the on-line measurement deviates considerably from the laboratory value, the characteristic curve is updated for better correspondence. Unfortunately, the updates are made in a rather haphazard manner. Recently, methods to systemise the updating (Latva-Käyrä and Ritala, 2005) or for detecting the need for proper determination of the characteristic curve (Latva-Käyrä and Ritala, 2005) have been presented.

The validity of the models is monitored similarly to validity of online sensors: comparing laboratory measurements and model predictions. This is because most simulation models can be considered as multi-input soft sensors with rather complex characteristic curves. The dynamic model validation is in its beginning, in particular, for models with considerable nonlinearity.

As available data is extensive, the probability of a data set containing false or missing values is rather high, although the reliability of the sensors was high. For example, if the reliability of sensors were 99%, the probability that a set of 1000 measurements would not contain false values at any given time is only 8%. Therefore

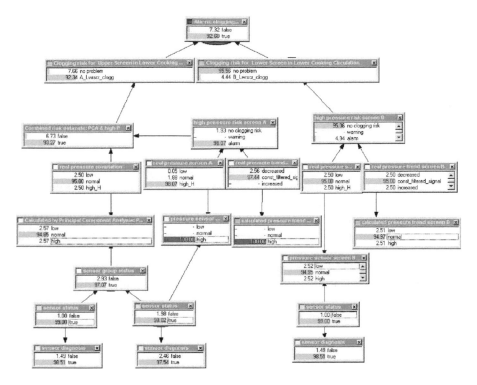

FIGURE 12.4 The Bayesian Network of the Danish company Hugin. (With permission.)

any extensive data analysis or simulation system must be able to detect false values and then be able to provide the users with analysis results and predictions with such incomplete data.

12.3   HUMAN SOFTWARE AND MACHINE INTERACTION NEEDS

The new functions will be implemented on a larger scale successively after extensive testing at different sites. A common issue for success is that the operators understand and accept the functions. This means that the functions must be robust and reliable, and tuning and maintenance also must be easy to do by the operator themselves, automated, or applied from remote by experts from the supplier, or a combination of these in a structured, well-defined way.

To make the functions understandable they need to be explained in a visual way with intuitive graphics and text support. How to do this is a key task for success. Researchers at universities, in industry, and at the suppliers have to cooperate in different ways by developing ideas and then test how these are accepted by the operators. We also need good documentation and help functions for the engineers who support the functions. It is not self-evident that they will easily understand complex functions, and especially not some time after the functions were developed.

So as a conclusion, this development work needs to be performed as a complement to the function development. If it is successful, we can foresee a strong impact on the profitability of the industry—which, of course, is the major driving force for the development.

12.4   RESEARCH NEEDS

One thread of current activities to be intensified in the future is following the objective of predicting and controlling paper properties based on process variables. All approaches are generally data based. Specific outcomes currently available are several soft sensors and model-based control loops operated in paper mills as described earlier. These projects accumulate complex knowledge about the way in which the process has an impact on product quality. To date these projects are individual approaches. An overview and a holistic scientific approach are still missing, these being future tasks for research.

A second thread is dealing with the calculation of paper properties based on the properties of the components defined by the recipe, such as pulp, fillers, and chemical additives. In this field most approaches are fairly generic and include deterministic models. Capabilities of available solutions are very limited. Clearly this is a research area in which a high number of projects are required to cover ground unknown today. One hindering factor for an open scientific exchange in this field seems to be that any breakthrough would give any industrial or scientific player a big edge over competitors. At the same time, despite high investments, none of the research teams has reached a satisfactory level of results within the past 20 years. This in the end could be an impetus towards collaboration in future research.

Looking farther into the future, the main objective of research will then be to interweave these two threads of process- and product-related research into a common web of knowledge. Deterministic process descriptions are currently being used and further developed. As a third strand of research activities, these could help to further improve the applicability of the knowledge gained.

As pulp and paper is based on natural materials processed in a complex way, another task will be to cope with random effects in order to minimise tolerances in mill-level operations, stabilising both product quality and the process. To do this it is required to interpret process data in a statistical context, to do further research to interpret the margin errors, and to more clearly understand process limitations. By combining mathematical models and lab tests, the way from lab to full-scale implementation can be significantly shortened and the possibility to shift between different qualities rapidly increased.

The focus of another area of research will be to describe and optimise the complex pulp and paper value chain on a model-based approach. As a side effect this will bring along the integration of process models with cost models. In the end, this could even prove to be a fourth thread, complementary to the ones described above. The task of future research would be to unveil currently-hidden factors that have an impact on the economic balance of the individual mill as well to interpret the interactions on the socioeconomic and macroeconomic level.

12.5   CONCLUSIONS

Optimal mill operations result in the optimal use of resources and in an increase of productivity, which means substantial financial benefits. While existing control methods are mainly based on staff experience, models providing better predictability and controllability of the processes would be of considerable value. Therefore, modelling and simulation are important tools to reach the two primary goals of the pulp and paper industry: the decrease of production costs and the increase of the product-added value.

During the past few years many applications for improving the process performance based on controlling different process variables have been implemented at an industrial level; however, the development of models to predict and optimise product properties based on process variables is still in progress. The main difficulty is that papermaking is a highly dynamic process and, in particular, the dynamics of the wet end are still not fully understood. Looking into the future of a knowledge-based process it seems that the interest in model-based control and optimisation will increase, modelling and optimisation of paper properties and model-based paper design will have a very high increase, while process simulation and off-line optimisation will remain the same from the production point of view. However, the equipment and chemical suppliers and consultants will increase the use of these tools to develop the processes and control strategies.

REFERENCES AND FURTHER READING

Alonso, A., Blanco, A., and Negro, C. (2005). *Current Use of Software in COST Action E36.* COST Report. Available at: www.costE36.org.

Andersson, N., Wilson, D., and Germgard, U. (2003). An Improved Kinetic Model Structure for Softwood Kraft Cooking. *The Nordic Pulp and Paper Research Journal* 18, 2, 200–209.

Blanco, A., Alonso, A., Negro, C., and SanPio, I. (2005). *Advanced Data Treatment to Improve Quality in a Newsprint Machine.* International Symposium IPE-PTS-CTP, Barcelona, Spain, 18–20 May.

Blanco, A., Fuente, E., Negro, C., and Tijero, J. (2002). Flocculation Monitoring: Focused Beam Reflectance Measurement as a Measurement Tool. *Canadian Journal of Chemical Engineering* 80, 4, 734–40.

Bogo, A. (2004). *How IDEAS Helped Manage Our Project Risk.* IDEAS User Group Conference (UGC), 4–5 November.

Box, G., Hunter, W.G., and Hunter, J.S. (1978). *Statistics for Experimenters.* New York: John Wiley & Sons.

Bryant, P., Samuelsson, Å., and Basta, J. (2003). Minimizing $BaSO_4$ Scale Formation in ECF bleach plants. *Tappi Journal* 2, 3, 3–7.

CO-LaN. (2005). www.colan.org.

COST E36. (2005). www.costE36.org.

Dahlquist, Erik. (2008). *The Use of Modelling and Simulation in the Pulp and Paper Industry.* EU Cost E36 Publication. Sweden: Mälardalen University Press.

Dahlquist, E., Ekwall, H., Lindberg, J., Sundström, S., Liljenberg, T., and Backa, S. (1999). *On-line Characterization of Pulp-Stock Preparation Department.* Stockholm: SPCI.

Dahlquist, E., Wallin, F., and Dhak, J. (2004). *Experiences of On-line and Off-line Simulation in the Pulp and Paper Industry.* PTS-symposium in Munich, 8–9 March.

Goedsche, F., and Bienert, C. (2002), *On-line Prozessanalyse und Computer Simulation von Herstellungprozessen.* Project report 37.

Gustafsson, J.E., and Kaul, V. (2001). A General Model of Deformation and Flow in Wet Fibre Webs under Compression. *Nordic Pulp and Paper Research Journal* 16, 2, 149–55.

Hakanen J., Mäkelä, M., Miettinen, K., and Manninen, J. (2004). *On Interactive Multi-objective Optimization with NIMBUS in Chemical Process Design,* MCDM, Whistler, BC. Canada, 6–11 August.

Hämäläinen, J., and Tarvainen P. (2002). CFD-optimized Headbox Flows. *Pulp and Paper Canada* 103, 1, 39–41.

Hauge, T.A., Slora, R., and Lie, B. (2005). Application and Roll-out of Infinite Horizon MPC Employing a Non-linear Mechanistic Model to Paper Machines. *Journal of Process Control* 15, 2, 201–13.

Honkalampi, P., and Kataja, M. (2002). Dry Content Analysis in Wet Pressing: Sensitivity to Pressing Variables. *Nordic Pulp and Paper Research Journal* 17, 3, 319–25.

Hutter, A., and Kappen, J. (2004). Analyse von Wasserkreisläufen. In J. Kappen and D. Pauly (eds.), *Wasserkreisläufe in der Papiererzeugung: Verfahrenstechnik und Mikrobiologie.* Munich: Papiertechnische Stiftung.

Ihalainen, H., Konkarikoski, K., and Ritala, R. (2005). Dynamic Validation of On-line Measurements: A Probabilistic Analysis. *Measurement* 39, 4, 335–51.

Jansson, J., and Dahlquist, E. (2004). *Model-based Control and Optimization in the Pulp Industry.* SIMS Conference 2004, Copenhagen, 23–24 September.

Kappen J., Hamann, L., and Cordier, O. (2004). *Identifying and Eliminating Stickies Problems When Using Recovered Papers.* Third CTP/PTS Packaging Paper and Board Recycling International Symposium, Grenoble, France.

Kappen J., Mueller, B., and Kamml, G. (2004). *Utilisation of Waste Heat—One Essential Approach of the Carbon Dioxide Problem.* The Finnish Paper Engineers' Association, PulPaper 2004 Conference: Energy, Coating, Efficiency, Helsinki.

Klemola, K., and Turunen, I. (2001). *State of Modelling and Simulation in the Finish Process Industry.* Universities and Research Centres, TEKES, Technology Review 107/2001, Helsinki, Finland.

Kokko, T. (2002). *Development of Papermaking Process Based on Integrated Process and Control Design.* PhD Thesis, Tampere University of Technology, Finland.

Laperriere, L., and Wasik, L. (2002). Implementing Optimisation with Process Simulation. *Tappi Journal* 85, 6, 7–12.

Lappalainen, J., Myller, T., Vehviläinen, O., Tuuri, S., and Juslin, K. (2003). Enhancing Grade Changes Using Dynamic Simulation. *Tappi Journal Online Exclusive* 2, 12.

Latva-Käyrä, K., and Ritala, R. (2005). *Sensor Diagnostics Based on Dynamic Characteristic Curve Estimation.* Tenth International Conference on Technical Diagnostics, Budapest, Hungary, 9–10 June.

Masmoudi, R.A. (1999). Rapid Prediction of Effluent Biochemical Oxygen Demand for Improved Environmental Control. *Tappi Journal* 82, 10, 111–19.

Mathworks. (2002). *Matlab 6.5 R13, User's Guide.* Mathworks Inc.

Mathworks. (2005). www.mathworks.com.

Mayböck, R. (2004). *The DLL Wrapper: The Andritz Vision for Advanced Process Control.* IDEAS User Group Conference (UGC), 4–5 November.

Miyanishi, T., and Shimada, H. (1998). Using Neural Networks to Diagnose Web Breaks on a Newsprint Paper Machine. *Tappi Journal* 81, 9, 163–70.

Nayar, Narinder. (1996). Workplace Ergonomics and Simulation. *Assembly Automation* 16, 1, 25–28.

Negro, C., Fuente, E., Blanco, A., and Tijero, J. (2005). Flocculation Mechanism Induced by Phenolic Resin/PEO and Floc Properties. *American Institute of Chemical Engineers Journal* 51, 3, 1022–31.

Orccotoma, J.A., Paris, J., Perrier, M., and Roche, A.A. (1997). Dynamics of White Water Networks during Web Breaks. *Tappi Jounal* 80, 12, 101–10.

Parrilla, R., and Castellanos, J.A. (2003). Aplicación del Control Predictivo en la Parte Húmeda de la Máquina de Papel. *El Papel* 108, 11, 63–67.

Räsänen, E. (2003). Modelling Ion Exchange and Flow in Pulp Suspensions. *VTT Publications* 495, VTT Technical Research Centre of Finland.

Ritala, R., Belle, J., Holmström, K., Ihalainen, H., Ruiz, J., Suojärvi, M., and Tienari, M. (2004). Operations Decision Support Based on Dynamic Simulation and Optimization, *Proceedings—PulPaper 2004 Conference–Energy, Coating, Efficiency, the Finnish Paper Engineers´ Association PI*, 1–3 June, Helsinki, Finland, 55–62.

Scott, W., and Shirt, R. (2001), Potential Application of Predictive Tensile Strength Models in Paper Manufacture: Part II–Integration of a Tensile Strength Model with a Dynamic Paper Machine Material Balance Simulation. *TAPPI Papermakers Conference*, 879–87.

Sundquist, A., Jakara, J., Aksela, R., Kaijaluoto, S., Pajarre, R., and Penttila, K. (2004). Modelling the Chemical Effects of Cumulating Metal Content in Bleach Plants with Closed Water Circulation. *Proceedings—PulPaper 2004 Conference, Energy, Coating, Efficiency*, Helsinki, 129–33.

Thomas, D.N., Judd, S.J., and Fawcett, N. (1999). Flocculation Modelling: A Review. *Water Research* 33, 7, 1579–92.

Vänni, P., and Launonen, U. (2005). Maximising Total Profitability with Dynamic Production Planning. *Appita Journal* 58, 3, 177–79.

Ylen, J.P. (2001). *Measuring, Modelling and Controlling the pH Value and the Dynamic Chemical State*. PhD thesis, Helsinki University of Technology, Espoo, Finland.

Williamson, Mark. (2005). Paperloop.com.

13 System Design
An Overview

Toni Ivergård and Brian Hunt

CONTENTS

13.1 INTRODUCTION

In this section we first consider the development of computerised process control systems. We then discuss some of the problems that the operator of an automated system may experience. Next we describe the advantages of involving the users in the planning process and offer some suggestions for project organisation. A more advanced form of participation in design is action research (AR). This methodology is discussed in a later part of this chapter.

It is clearly important for the planning and design engineering of new control rooms to be carried out in stages. This process should preferably be seen as a continuous ongoing task in the organisation. In such a process, the extent to which changes are made varies. Sometimes the changes are small and incremental; at other times, the changes many be large, more comprehensive, and revolutionary. Presented here are a methodology and a philosophy for the development of process control systems and control rooms. The most important points in this philosophy are:

1. The work must start with an unprejudiced and unconditional analysis of goals and functional requirements.
2. Planning for personnel in different job positions must be done in parallel with the technical planning.

3. A systematic method must be used to determine which functions should be done by machine and which by human operators in order to achieve the given aims.
4. Great emphasis should be placed on the design of the interface between the machine and the human operator.

Figure 13.1 shows a schema of a project planning system, which includes the roles of the human factors specialist.

13.2 DESIGNING NEW SYSTEMS

13.2.1 THE TRADITIONAL APPROACH

When designing new systems, factories, and products, it is usual to follow the procedure shown in Figure 13.1(a). The designers usually first determine the aim or goal of the system and then decide its technical design. This process is often done in several stages. The overall principal design comes first, followed by a more detailed technical design process. This process produces finished drawings and specifications. Construction can then get under way. Usually during the construction stage, the organisation begins to seek new staff, conducts recruitment activities, and hires new staff as appropriate. Existing staff may also be retrained where staff relocation or transfer to a new system is involved.

Problems occur most often when personnel begin to use the new system. Some people may need further training and various types of instruction and others may be incapable of using the new system. If a large number of personnel cannot use the technology to do the work it may be necessary to alter the technology. Another common problem area concerns environmental issues. It is not unusual for issues in this area to be overlooked. Consequently, environmental requirements may not be fulfilled, and technical measures must also be taken.

13.2.2 SYSTEMS ANALYSIS

The traditional process is spread over several stages, is ineffective, and often gives poor results (see Figure 13.1b). This construction design is sometimes referred to as the 'relay race' model, referring to the stages that are as sequential as the passing of the baton in a relay race. One alternative is the 'rugby scrum' approach, where the expert teams collaborate to execute several stages in parallel, as shown in Figure 13.1b. We first present an overview of Figure 13.1b, and then describe each of its subelements in more detail. Figure 13.1b should be regarded first as a philosophy of work; it cannot be treated as a concrete and general work model that can be applied directly to different circumstances.

The goal of a project is defined in accordance with traditional practice and defined in terms that are as concrete as possible. This is followed by a very important element, namely an unprejudiced description of the functions required for the system to fulfil its goal. It is important for the function analysis to be as detailed as possible. It must, however, be phrased partly in abstract terms, because in the next stage decisions will be made in the way in which the various functions will be

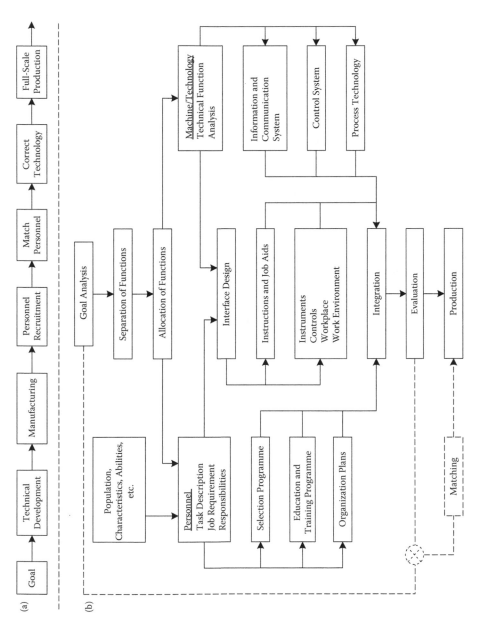

FIGURE 13.1 Project planning system and the role of the human factors specialist.

carried out—and whether these are done by a human operator or a machine. The allocation of functions to human or technical components is done in such a way as to fulfil the various criteria regarding health, safety, and comfort. This is a very important element in the planning work, enabling the best possible system to be obtained. The degree of automation in the final system, for example, is determined through allocation of functions in this stage, and the form of technology to be selected is also determined at this point. Allocation of functions indicates which tasks in the system will be done by the human operators, in accordance with the left-hand side of Figure 13.1b. Descriptions of the different tasks involved determine which job and craft skills are required for each one.

Starting with the descriptions of the tasks and job demands, a suitable allocation of tasks is made to the various posts. These in turn will then form the basis for the design of the education and training programmes, organisational plans, instructions, and different types of job aids. In certain cases, and for particular systems, it may also be necessary to have special selection programmes. The technical design work is carried out in parallel. First, the designers decide upon the technology to be used in the process itself, then which control systems are to be used, and how the information and communication systems are to be designed.

At the same time as these two activities are going on, there is close cooperation between the design of the various personnel programmes and the design of the various technical programmes. This collaboration results in the design of the area of contact, or interface, between personnel and technology—that is, the design of instrument and control workplaces (such as work surfaces, tables, and chairs). This work normally uncovers certain unsuitable or incorrect allocations of functions. For example, it may be discovered that functions which are seen to be difficult to carry out or unsuitable have been inappropriately allocated to personnel. These need to be transferred to the technical side, and a technical solution for the functions sought. Alternatively, functions for which it is difficult to find technical solutions may have been allocated to machines. In such cases these are transferred back to the personnel side. In this way, the technical and the personnel sides of the work operations can be worked through and both systems brought together. After this, it may be apparent that there are shortcomings in the degree to which the goals are fulfilled. In such cases, various corrective actions must be made. By following this process, the need for corrections at a later stage will be reduced.

It is important to note that if environmental factors are taken into account from the beginning of the design process, most of the problems normally encountered can be avoided or resolved. This means that the planning must be carried out in accordance with the 'ergonomic system' model, shown in Figure 13.1b. Without such a systematic working method, the planning process cannot create a good working environment in the finished system. This model has also proven to be an advanced way to stimulate creativity and thereby discover new innovations and solutions to problems. Allocations of functions fulfil an important role in this process. At the same time, it has also proven to be a useful tool through which to structure and to optimise the balance between human operators and technology.

13.2.3 GOAL ANALYSIS: AN EXAMPLE

The importance of formulating the goals in as concrete a form as possible has already been emphasised. Analysis of the system's goals can best be done under the following three headings:

1. System Task
2. Criteria
3. Limitations

The *task* of a system may be to convert raw oil into various petrochemical products. It may also be to convert potential energy in the form of a stored mass of water into kinetic energy, and then to electrical power. A number of *criteria* apply to this task, for example, those of an economic nature. From the ergonomic viewpoint, the important criteria are to create a safe environment for the people in the production process and for those in the community. It is also important to create interesting and meaningful jobs as well as comfort and well-being for those involved.

Limitations are factors that are constrained by the environment and society as a whole. Examples of limitations are laws and regulations. The context in which the system is to function also gives rise to a number of limitations and conditions. Before a system is developed, therefore, it is important to determine how the system can function together with other systems (which can also affect the system in question).

A system for converting water storage energy into electrical power is, of course, dependent on which other types of power production it will be working in conjunction with. It is also dependent on how the electrical power is to be distributed. One predetermined background factor is also the design of the water system. By taking these various parameters into account, a number of factors that are important to the system itself may be enumerated, together with all the limitations on the system.

13.2.4 FUNCTION ANALYSIS

Function analysis often leads to a form of function diagram. One basic rule in this work is that it should be free from preconceptions and not bound by any special components or solutions worked out previously. As an example, consider the transport of goods through a supermarket checkout. The question is whether this should be done mechanically or manually with the aid of a conveyor belt for use by the customer or the cashier. This choice should only be made at the function allocation stage, and then only on the basis of certain given criteria (those produced at the goal-setting stage and ergonomic factors are of primary interest in this case). The function analysis must be achieved without prejudice as this is the basis for the resultant allocations being the optimum solutions. The unprejudiced function analysis can then help to create new solutions that are better for mankind.

Thinking without preconceptions is difficult, as it involves having to think in abstractions. When we think, we generally prefer to be concrete. Abstract thought allows us to think about the things we are used to seeing or noticing from a new perspective. If we consider the cleaning of textiles we immediately think of the physical objects that we connect with this task, such as the washing machine, spin drier, or a person in a specific role. But we do not necessarily have to have these components to fulfil the given task in the way stated in the goals. Figure 13.2 shows a simplified function for the task of cleaning textiles.

The various subfunctions needed for the execution of a particular task can each be treated as subtasks with their own criteria and limitations. In this way it becomes easier to think in abstractions, as the tasks and goals are expressed in abstract terms.

In general, it may be said that function analysis should be carried out in as much detail as possible. From the ergonomic point of view, however, it is not desirable to go further than the level determined by the human operator's inherent method of functioning.

Let us take the hypothetical case where all functions to be defined must be able to be carried out by people (if this is thought to be desirable in the subsequent function allocation process). It will not then be necessary for these functions to be expressed in such detail that it is possible to say beforehand that several different subfunctions must necessarily belong together, for example, for physiological reasons. The visual centre and decision centres in the human brain are, of course, connected. It is therefore meaningless to separate these functions. If it is decided to let the person carry out a visual function, such as identifying the price on an item, the subsequent decision functions (for example, which keys are to be pressed) must also be carried out by a person.

It may be desirable from the technical point of view to carry out a more detailed function analysis after the first allocation of functions. This is because during the choice between the various technical solutions it may be necessary to go into considerably more detail.

13.2.5 ALLOCATION OF FUNCTIONS

Allocation of functions can be done in more or less detail. If ergonomically-orientated function allocation is used, it is relatively simple to specify the level of detail at which the functions allocation should be carried out. In the function allocation stage that follows function analysis, the functions are allocated between humans and machines/personnel technology. In this way, one can specify which functions are to be performed by the human element and which by a technical component. It is therefore not meaningful for a function allocation to be written in terms that are too abstract. This can result in all the functions following function analysis needing a combination of both human and technical solutions.

An allocation of functions according to Figure 13.2 is thus far too general. It probably requires both human and technical work in the form of controlling and at the same time it requires technical equipment for water. One should therefore

FIGURE 13.2   Simplified system for cleaning textiles.

divide up the functions into considerably smaller units. Ergonomic allocation of functions is concerned primarily with the distribution of functions between human and machine. This later stage is the first time that a further allocation is carried out, between the different technical solutions on the one hand, and the allocation of different tasks between different job holders on the other. In the textile cleaning system discussed above, therefore, the question is then asked whether the sorting of dirty textiles should be done with technical aids, and whether feeding-in should be done by hand or by technical means. The various sections in the process can be covered in the same way.

Function allocation is carried out on the basis of certain criteria. In the United States, various forms of productivity criteria are used in function allocation in the development of technical systems and methods for product development. The attempt is made to determine which is the more efficient at carrying out a particular function, a human or a machine. Special tables have been worked out that allow one to read off when man is more suitable or when one should use a machine. One such table is the well-known 'Fitts' List', which is reproduced in Table 13.1. This narrow selections method is much too limited today to fulfil any practical function. Such efficiency criteria are, of course, important, but considerably more detail than that given in Fitts' list is required. However, Fitts' list can be used for a rough classification.

It is also important to take economic criteria into account but this is beyond the scope of this book. The most important factor we are concerned with is to create meaningful work that does not cause injury or fatigue to people; that is, we are concerned with the various forms of human criteria. The criteria should be decided at the same time as defining the system goal. The criteria are used in the evaluation of the human versus the technical alternative for the performance of different functions. We now look at some examples of how this form of discussion may be carried out in an allocation of functions.

Machines are often best at performing certain types of physical activities. This is particularly true of very simple operations with no fine manipulation, and where the work is monotonous and repetitive. The machine is almost always to be preferred over human beings for heavy physical lifting tasks, both from the efficiency and the human viewpoints. The technical solution, however, can be very costly if operations involve more complex and fine manipulations. In our example of textile cleaning, the feeding-in, unloading, and transport tasks should preferably be done mechanically, especially if the system is also to be suitable for an older population.

The human sense organs are in all cases superior to technical equivalents, particularly if the cost aspect is taken into account. In addition, different types of control work—where the person needs to use vision and hearing, for example—usually also involve important operations that people prefer to control themselves. For certain simpler types of sensory work, however, technical aids can be the best alternative. For simpler decision making, it is relatively easy and cheap today to use technical devices as an alternative to human inputs. But even here it is important to remember that human beings get satisfaction from controlling a process. In addition, a human being has considerable advantages when it comes to making simple decisions. For example, a person can make small adjustments to the decision process without having to use programmed information. A washing drum can be allowed to rotate a few

TABLE 13.1

Relative Advantage of Man and Machines

Characteristic	Machine	Man
Speed	Outstanding	Soon slows down
Power	Constant and large	Maximum of a few h.p. for a few seconds
Repetitive ability	Ideal for routine and repetitive jobs	Unsuitable, should be done by machine
Complex activities	Multichannel, can do several things at once	In general single channeled (except for kinaesthetic information)
Memory	Large short-term memory and suitable for exact reproduction	Good long-term memory, suitable for principles and strategies
Decision-making	Good at drawing simple conclusions from large amounts of data (deduction)	Good at drawing complex decisions based on small amounts of data (induction)
Calculation	Fast and accurate. Bad at correcting own faults	Slow and inaccurate. Good at correcting own faults
Input sensitivity	Some types of stimulus man cannot sense, e.g., radioactivity, light	Large amplitude range. Ability to determine several quantities from one stimulus, e.g., the eye luminance
	Poor pattern recognition	Very good pattern recognition. Can extract information against very noisy backgrounds
Intelligence	None	Can deal adequately with unforeseen and unpredictable situations
Manipulation ability	Rigid (only for certain predetermined tasks)	Flexible (can carry out many different types of tasks)

Source: Modified from Fitts and Posner (1967).

minutes longer if the human operator sees (from sensory inputs) this as desirable, for example, due to the quality of the textiles being processed. It would be very costly to get a technical device to make such small modifications automatically. In complex decision making in systems with high inertia (for example, manoeuvring large ships, administrating parts of financial systems, or energy systems), operators obtain great benefits from using simulators and mathematical modelling of systems.

In general, it may be stated that a human being is a very poor monitor, and therefore should not be given the job of monitoring technical systems. On the other hand, a machine often makes a tireless monitoring device.

A good allocation of functions between humans and machines for the job of cleaning textiles may be as follows. The operator sorts the textiles according to quality and dirtiness and puts them into different bins. The machine is preprogrammed to handle the material from the different bins in different ways. Depending on the type of textile and type of water, the machine itself regulates the temperature, water quantity, and type and quantity of the various cleaning chemicals. After the textiles have been processed with water and chemical additives and centrifuged to remove excess moisture, they are transported down into a drum for drying. Removal of the dried textiles is done manually.

The different forms of function allocation can be discussed in a preliminary stage in this way. One can then analyse the different alternatives in more detail in order to find the 'optimal' solution. It is clear that, for the parameters in this case, it is very difficult to make any sort of quantification that would allow an exact form of optimisation procedure to be produced. This may be possible in certain cases, while in others it is considerably more difficult. Efficiency criteria can often be viewed in economic terms and thus allow a certain degree of comparability with other criteria. The result is a quantifiable evaluation system.

It is considerably more difficult to measure the human criteria (in terms of comfort, fatigue, and accident risks, for example) in such a way as to be comparable with other criteria. The difficulty of taking account of the human criteria in a more quantifiable way during function allocation must not result in their being dismissed altogether, as these aspects are often of the utmost importance. Within the field of technical production research, an attempt is made to translate the various forms of working efficiency into economic terms. However, it is not generally an easy method, and it is important to create a method in the future where other than just the economic criteria are taken into account. Only under these circumstances will we be able to build up a technology that is suited to people. In other words, it is important that a methodology for product development is used that allows us to steer the technology in the required direction, whilst also taking full account of what is best for the operator.

13.3   HOLISTIC VIEW OF PLANNING

A starting point for the development of computer software should always be to obtain an understanding of the stereotypical expectations of the operators. There are, however, critical difficulties with this method. Different individuals have different mental models and therefore different stereotyped expectations. Thus, it is not relevant for the software designer to use himself as a reference in the prediction of plausible mental models in the mind of the intended operators. Instead the designer needs to refer to his basic knowledge in perceptual, cognitive, and social psychology using people who are as similar as possible to the real users of the system.

To obtain a structure, an analysis of the manual and mental tasks involved must be carried out. These tasks (functions) in turn comprise many different subtasks (subfunctions) that also have to be described. When this has been accomplished, one has to carry out an allocation of the tasks to the operator and to the computer. The operator should have a general understanding (not a detailed knowledge) about the tasks allocated to the computer and the compatibility of the operator's mental model related to this situation. Two different complementary approaches are used for this process: the systematic, and the holistic and participative. The *systematic approach* is explained in Section 13.2.2 (Systems Analysis).

Traditionally, dialogues are designed last in the process of developing a new computer system. This is not advisable. The dialogue design should be regarded as a design process of its own party separated from the total system development and initiated from the outset. The result of the dialogue design should be a specification to be used in the rest of the system development.

The *holistic approach* implies that the design work is carried out to some extent on an *ad hoc* basis. This form of design process is claimed to give room for more creativity but is, of course, less predictable and might be difficult to use for very large projects. The holistic approach often also implies that one is working in close cooperation with the intended users of the system. *Participative design* is especially important in the design of dialogues. Working with the users would give access to the users' knowledge and expertise about their own working conditions. This will help us to carry out the task analysis and to define the operator's 'mental models'. To include the operators in the development will also increase their motivation and willingness to cooperate in a positive and constructive way in the implementation of the system. The operators can also be used as subjects in different forms of expert evaluations. Expert evaluations can be done on either the whole or on parts of the dialogue.

It is, however, often difficult to involve users in a traditional process of system design. The users normally do not have sufficient levels of programming skills nor are they used to abstract thinking. The design procedure has, therefore, to be changed so that instead of using different theoretical development tools, it can be visualised in a more practical and realistic form. At a very early stage of the design, one can use 'mock-ups' of the dialogues. These mock-ups could simply be paper-and-pencil exercises that describe and illustrate the different dialogues. Another alternative might be to present the different dialogues on a personal computer. However, this form of dialogue consists only of visual illustrations and does not allow any real-time interaction.

The next refinement could be the use of prototypes. The prototype is a simplified version of the final system developed on a personal computer. The prototype will allow some real-time interaction between the system and the operator but it does not cover, or does not incorporate, the use of a complete database.

13.4   ORGANISATION OF THE DESIGN PROCESS

Recently, there has been considerable emphasis on the rights of users to have an input into product/system development. To achieve an effective input from employees, it is necessary to establish an organisation for the planning work as a complement to the planning philosophy presented in Figure 13.1. This organisation is probably best determined in the company's safety committee. Figure 13.3 shows one suggestion for how such an organisation might be set up for the development of a new, large system such as a new control system for a process.

A central steering group is needed for the project within the company. The steering group would include the safety committee and anyone in the company who has responsibility for long-term planning. The central steering group also includes a project leader who is directly responsible to it. There is, at the project leader's disposal, a project leader's group that includes representatives for the more important

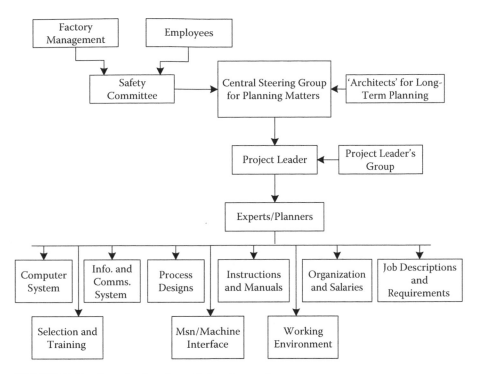

FIGURE 13.3 Organisation chart of planning work.

experts. This 'expert group' would include representatives of the operators affected and representatives for safety and company health functions, as well as the various technical specialists. The various experts and expert groups will work under the project leader. Among the experts and expert groups there should be special groups that deal with various personnel questions, for example, the production of job content analysis and job descriptions. Another group should be responsible for the selection and training programme, one for the instructions, manuals, and job aids, and one for organisational questions. If the project is a small one, some of these groups can be combined.

The central steering group decides upon and works out the goal analysis in conjunction with the project leader. Then the project leader, in coordination with the project leader's group, carries out a suitable allocation of functions. This group also produces a preliminary proposal for function allocation. The proposal is then sent for consideration through the various groups in the company such as the safety committee and company management. Only after this stage can the central steering group decide on the first suitable allocation of functions. Any substantial changes from the first proposal for allocation of functions must again be discussed by the central steering group, which will then determine whether the approval of the various groups is again required. Figure 13.3 shows an organisation chart focusing on planning work.

As time is often short when the actual planning work gets started, a high proportion of the responsibility for the first part of the work (that is, that concerned with personnel and technology) lies with the project leader and the project leader's group. Only the larger and more important matters are given to the central steering group. In order to have a last chance to assess whether the project shall be carried on to completion, a preliminary evaluation is made. This first major evaluation is made using a model or a pilot plant; if this is found to be impossible, drawings and progress charts can be used instead. The groups involved in dealing with personnel matters include employees' representatives. These matters are also of concern to the central steering group and in the project leader's group. It may also be desirable to include employees' representatives in certain of the more critical technical groups.

Ergonomic and/or industrial psychology expertise is needed, for example, in the group working on the interface between the personnel and the technological system and equipment. Such special expertise may also be required in the groups working on instructions and manuals, in the finalising of the job descriptions, and also possibly on the production of a selection and training programme.

13.5   JOB ORGANISATION AND ALTERNATIVE FORMS OF AUTOMATION

Advanced technology for control and information processing, often computerised, has infiltrated almost explosively into more and more areas. In spite of optimistic expectations, this has often led to deteriorating work conditions for the people involved. This has been described in a large number of ergonomic and social/psychological studies.

Within the process industries, it is found that the U-shaped curve expresses the relationship between work motivation on the one hand and technical development on the other (see Figure 13.4). It seems as if work motivation in the modern, computerised, highly-automated process industries is rather low, and in many cases is approaching that of 'assembly-line' jobs. It is also found (Figure 13.5) that variations in work load have become even greater and that the low-load periods have become longer and even lower. Automation has often led to a severe reduction in job content and an increased degree of outside control (due, among other factors, to the increase in the degree of continuity in process control) with the introduction of computer technology.

Figure 13.6 attempts to show in a highly simplified way how computerisation is traditionally carried out. Figure 13.6 is not just relevant to the process industry, but probably has considerable validity in general application. The reason for this is mainly that technical people working on automation development have hitherto worked fairly conventionally. As such, they have taken a much-too-limited view of people and their work requirements. The view was perhaps even mistaken. The starting point has been the old technology, what it knew and what it did. The developers considered the tasks the old technology could carry out and then determined the jobs of the human operators. Using this as a background, advanced computerised technology has been allowed to take over the work both of the conventional technology and of the person. One obvious consequence of this is that the content of people's jobs is drastically reduced.

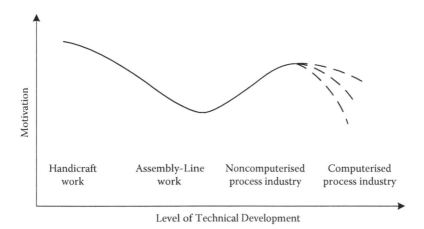

FIGURE 13.4   Level of motivation at different levels of technical development.

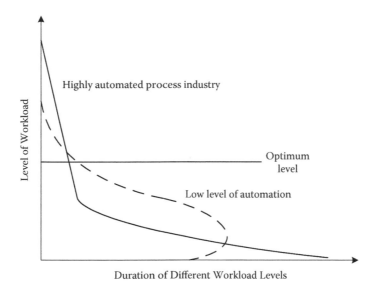

FIGURE 13.5   Relationship between workload and different levels of automation.

It is interesting to note in this connection that the new computerised technology has intrinsically more 'power', and the ability to do much more than could be done by the old system and the human operator. In order to control a technical process optimally, there are theoretically an infinite number of functions available. The human operator and traditional technology, with their limited abilities and resources, can only make use of a small part of these. Modern computer technology has the potential to do other tasks; it is therefore possible to utilise functions that were not

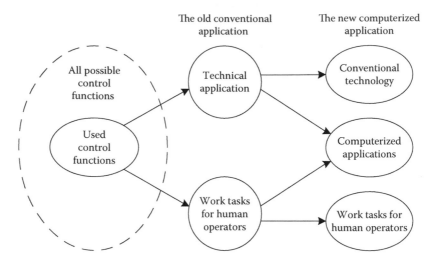

FIGURE 13.6   Conventional approach to the design of automation.

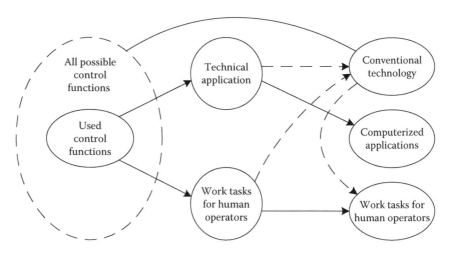

FIGURE 13.7   Alternative approach to the design of automation.

earlier used (see Figure 13.7). With a little imagination, new and unconventional solutions for controlling processes could thus be found.

In our research (Ivergård et al., 1982; Hunt, Burvall, and Ivergård, 2004) we have shown examples of how it has been possible, with the aid of computerisation, to give the operator refined information on the functions, strengths, and weaknesses of the process. New, advanced computer technology can give the operator better information, and at the same time new abilities in the controlling of the process. In this way, the operator can be included right in the centre of the process. Actually the operator is the most central function. The operator is given a decisive role in determining the production function and the production results in terms both of quality and speed. The operator should thus be able to become even more important than before.

In addition, the operator can be given the opportunity to build up new forms of job skill and also the possibility of reinforcing the old, traditional job skills based on his or her personal knowledge and experience of the process. In other words, the operator does not need to be a mechanical cog in the process but can take the central control function. With this form of alternative development, the job will become at least as interesting and meaningful as the previous work—that is, the job becomes of value.

Apart from making the design of the job interesting and meaningful for the individual, the operator can, with the aid of computerisation, get a better insight and overview of the whole process. The computerised system can provide an overview of the production results and economic aspects within the company, which increases the operator's ability to take a meaningful part in the overall decision-making processes in the company.

It is outside the scope of this short review to discuss in any more detail the background as to why this line of reasoning is actually true. Some of the reports published by Ergolab (Ivergård et al., 1982) give a somewhat better exposition of this background and the reasoning behind it. Comparative studies of the effects of computerisation within manufacturing industries have been carried out in Great Britain and Germany (Sorge et al., 1982). These studies also show that computerisation does not need to lead to any downgrading of jobs on the factory floor. Some related aspects about learning and creativity in control room work are also discussed in Chapter 11.

REFERENCES AND FURTHER READING

Archer, L.B. (1965). *Systematic Methods for Designers*. London: Design Council.
Bailey, R.W. (1982). *Human Performance Engineering: A Guide for System Designers*. New York: Prentice-Hall.
Govindaruju, Majorkumar, Pennathur, Arunkumar, and Mital, Anil. (2001). Quality Improvement in Manufacturing through Human Performance Enhancement. *Integrated Manufacturing Systems* 12, 5, 360–67.
Gagne, R. (ed.). (1962). *Psychological Principles in System Development*. New York: Holt, Rinehart and Winston.
Hughes, John, King, Val, Rodden, Tom, and Andersen, Hans. (1994). Moving Out from the Control Room: Ethnography in System Design. *Proceedings of the 1994 ACM Conference*, 429–39.
Hult, Lars, Irestig, Magnus, and Lundberg, Jonas. (2006). Design Perspectives. *Human-Computer Interaction* 21, 5–48.
Hunt, Brian, Burvall, Patrick, and Ivergård, Toni. (2004). Interactive Media for Learning (IML): Assuring Usability. *Education & Training*, 46, 6/7, 361–369.
Simpson, C.A., McCouley, M.E., Poland, E.F., Ruth, J.C., and Williges, B.H. (1985). Systems Design for Speech Recognition and Generation. *Human Factors* 27, 115–41.
Singleton, W.T. (1967). Systems Prototype and Its Design Problems. *Ergonomics* 10, 120–24.

Sorge, A., Hartman, G., Warner, M., and Nicholas, I. (1982). *Microelectronics and Manpower in Manufacturing*. Berlin: International Institute of Management.

Turney, Robin D. (2007). The Überlingen Mid-air Collision: Lessons for the Management of Control Rooms in the Process Industries. *Loss Prevention Bulletin* 196, 3–7.

Part VI

Conclusions

14 Conclusions and Recommendations in Summary

Toni Ivergård and Brian Hunt

CONTENTS

14.1 OLD KNOWLEDGE AND SLOW IMPLEMENTATIONS

It is very difficult to change people to fit machines, and it is of course not desirable to do so. After all, machines are made to serve humanity and not vice versa. It is therefore better to design the machine from the very beginning to fit the human user. That this does not happen more often depends largely on the fact that many engineers lack factual knowledge of human characteristics. There is a need for a new type of engineer with a basic knowledge of the human being—in other words, one with knowledge of the relevant aspects of psychology, physiology, anatomy, and a certain amount of medicine. An engineer with this background is known as an *ergonomist*.

The concept of ergonomics was introduced in 1949 as a title for the inter-disciplinary teamwork between various human scientists. It was started by a group of British scientists, primarily psychologists and physiologists, and the originator of the name was the marine psychologist K. F. H. Murrell (see Murrell, 1967). The word is derived from the Greek *ergon*, meaning work, and *nomos*, whose nearest meaning is natural law or system. The conjunction of ergon/nomos was also used in classical Greek times for a workers' protection law, probably the world's first Health and Safety at Work Act. The Ergonomics Society was founded in 1949. The journal *Ergonomics* was founded in 1957. In the 1960s there was a fast growing interest in the physical aspects of ergonomics (for example, the physical design of workplaces, such as chairs, tables, or desks) and the physical environment (such as noise and acoustics, thermal visual conditions, and lighting design). During the 1970s and 1980s there was a rapid increase in research and development in the social and psychological perspectives of people at work, and the term *information ergonomics* was coined (see Ivergård, 1982). During the 1980s to the end of the 1990s there was a change of focus from human beings at work to job and organisational design, knowledge, and the human/intellectual capital, including issues of learning at work. From the year 2000 macro aspects of work and knowledge come in focus (a new kind of labour economics). Richard Florida (for example, 2004, 2005) and others (for example, Ivergård and Hunt, 2007) claim that creativity will become the most demanded skill and valued 'class' of talented people. Discussions about the importance of participation, involvement, and empowerment were reinvented. As in this handbook, the need for holistic and broader approaches is in focus. This need involves a combination of real work practices, while at the same time concerns itself with individual learning and developing as well as an individual's need to be a creative contributor to the development of his or her own work, workplace, and firm.

The old edition of this handbook was published in 1989. To a very large extent it was a product of the development and research during the period from 1970 to the end of 1980. At that time a new and much deeper understanding of how human intellectual and perceptual capacity could be used in control room and systems design. This new edition of the handbook aims to integrate the new perspectives from the end of 1980 to 2008. But we can also notice from our studies of new control rooms/centres that the knowledge of the old edition still has great relevance to 2008.

14.2   THE HUMAN USER AS A COMPONENT

The degree of success when planning human/machine systems, and in particular monitoring and control systems, is entirely dependent upon being able to predict the consequences of different design solutions. The ability to predict the consequences of a specific solution is dependent in turn upon the designers' insight, knowledge, and experience of the various components that influence the final function. In a human/machine system—that is, in every system where people work—it is essential to consider people as a significant cognitive, emotional, and feeling part of the system.

In certain cases, solutions can be found through common sense, but when planning more complex human/machine systems, common sense alone is seldom adequate. Instead, it may create completely the opposite effect. When planning new, advanced technical solutions, interaction with people is not always easy to predict. This is in total contrast with day-to-day situations, where our daily experience provides us with much more information.

In something of a cookbook style, we have attempted to provide the majority of basic recommendations about people as a component in complex human/machine systems. The ability to predict the final functionality of the system will improve by using the knowledge presented in this book. This should lead to a better working system. In this context 'better working' means a system where the final result, the final product, is that which was originally intended. In many instances, great demands have been placed upon factors such as production safety, product quality, and general safety. The possibility of achieving these goals will hopefully be increased by applying the recommendations presented in this book.

There are still people who, when planning, do not believe in the necessity of systematically and seriously taking into account the people who will work with and use the machine and its technology. There are those who believe that a human being as an individual can be taught to fulfil and to adapt to all types of behavioural criteria. This is incorrect. It is important to understand that the human is a highly reliable system component and functions according to predetermined 'laws'. This reliability is considerably superior to the reliability of the technical components. Let us illustrate this point.

A tram quite suddenly makes an unscheduled stop at a very busy crossroads; for what purpose? The least likely answer is that the operator—the driver—left the tram to pick flowers. Alternatively, it is highly likely that a technical problem has arisen. The probability of total breakdown of the human component is considerably less than that of corresponding technical parts. On the other hand, humans display many variations and smaller deviations. This is characteristic of the human factor. This is also one of the reasons why errors are attributed, often unfairly, to people rather than technology, which is often the real cause. As technology is often poorly adapted to the individual, the human (with his or her prerequisites and qualifications) has difficulty in achieving certain given performance demands. One often hears about 'human error', but these types of errors in fact originate from poor technological solutions. These poor solutions in turn can be attributed to the person who has designed the system and this person's lack of knowledge in designing systems.

14.3   BUSINESS CONCEPTS AND SYSTEM DEVELOPMENT

There can be many and varied reasons behind planning a new process industry or a large reconstruction project. The underlying reasons are often associated with governmental industrial policies, employment policies, and so on. Normally there is also some form of business strategy based upon a business concept. If limited to development projects of new computerised control systems with an increased level of automation, the definition of goals often becomes more diffuse. When Ergolab carried out a series of investigations in Swedish process industries, the operators often questioned the motivation behind the introduction of new computerised systems. It was quite obvious from interviews carried out during the investigations that the reasons were rather diffuse and not particularly precise—nor, in the industries studied, had any effort been made to establish to what extent goals were fulfilled. This was partly due to goals being unclear, which led to the inability to follow up and measure goal achievement.

In order to succeed with computerisation it is, of course, a prerequisite to define what the new computerised control system is to achieve. Often manpower reduction is put forward as an important goal or subgoal for an increased level of automation. As emphasised earlier in this handbook (for example, Figure 1.2), reducing the number of employees is a questionable basis for automation. There is, of course, a relation between automation levels and staffing requirements, but it is equally obvious that there is a clear staffing minimum irrespective of the level of automation.

Staffing demands increase again once beyond a certain level of automation. Here, it is important to consider the entire picture and not just the manning of the control

room. High levels of automation can lead to increased demands on service and maintenance staff. Further, very high levels of automation demand extremely large investments in planning, programming, and other essential factors. The most important
and realistic reasons behind automation are those that concern increased production
safety, increased product quality, and the like. Further, significant improvements in
the working environment can be achieved when designing control automation. This
in turn can have positive secondary effects on factors such as job skills and staff
turnover. New technology is not of value *per se* and might only be the latest technical
'toys' to please the advanced engineers.

It is always important to follow the traditional planning stages as presented
below, even when planning a new control system for a process industry:

- Formulate a clear and well-defined *business concept*. The concept is
  achieved by a number of tangible and measurable goals.
- Define various strategies in order to fulfil these business goals.
- Specify those activities and processes and the type of equipment intended
  for procurement based upon the business strategies.
- Develop a human resource (HR) concept that supports the business con-
  cept. This should be developed at the same time as the business strategies.
  The HR concept should be tangible and defined in measurable terms.
- Formulate a number of HR *strategies* based upon the HR concept.

Human resource strategies form the basis for those activities, processes, and equipment found in the completed design. Further, HR strategies lead to other internal
personnel activities, for example, management training, leadership development,
systems for staff developments, operator training, and last but not least, the development of people (the users) into experts.

Only once these five stages have been thoroughly completed is it possible to
decide which functions should be automated and which should be carried out by
the human element. It is not possible to discuss suitable forms of automation before
knowing which goals should be achieved. However, once this information is available, decisions can be made about the way and level of investment in the human
factor, and it is here that a handbook of this type is an integral part of the planning
process. Some important conclusions and recommendations are summarised in the
following section.

14.4   SOME RECOMMENDATIONS IN SUMMARY

Work in control rooms can either be organised around maintenance, engineering/
field service, and error/fault handling, or around planning and process optimisation.
In the first case, work in the control room is in some way combined with maintenance
and service tasks carried out in the plant itself. In the other case, the control room
operator is given the task of carrying out more advanced planning and optimisation
work. Here, for example, computers are used to control and monitor the process and
also to carry out other tasks, such as simulating various process conditions. These
simulations can form the basis either for controlling the process, planning future

tasks, or replanning or redesigning the process. It can also be used for learning and development of the operators' skill and development, including tacit knowledge and mental models.

There are no simple rules about the manning of control rooms, for example, in relation to the number of monitored control units. In general, a pure on-line process without any elements of batch grouping is the most difficult to control. The characteristics of the work to be carried out in the control room depend partly on the type of process and partly on the type of staff available. The latter type of work—that is, planning and optimising—demands access to staff with a more theoretical education and training. If emphasis is placed on service and maintenance, control room operators can be recruited from plant employees. The optimal is often a combination of a rather advanced theoretical background with a good practical experience gained from in-the-plant maintenance work.

Many and varied types of visual display units (VDUs) are used in modern control rooms. Often this means great advantages, but one should not be overconfident about the large flexibility that visual display units offer. Further, there is nearly always a need for some form of general or overall display of the process. Sometimes it can even be of value to keep the old, more traditional type of control panel with both instruments and controls. An alternative would be to simulate some of the control settings on the overview display.

With the transition from traditional controls to keyboards, the modern, computerised control room is in danger of losing an important source of information. Classical control panels or modern simulated versions are often vital in order to achieve efficient production. Other aids, such as knobs, the light pen, joystick, mouse, and so forth can be of additional value. Colour displays should be used with some caution, and for valid purposes. Colour should not be the primary source of information, and the use of colour should be very clear and unambiguous. The use of voice control and speech (and other auditory) information can be of value, particularly for various forms of emergency situations.

Do not overload the visual senses of the human operator. Human beings have more senses than vision. In the next generation of control rooms we will see more use of all the other sense organs. Using the whole range of senses has the potential for the operators to improve their tacit knowledge and mental models of the system. And, in turn, this will improve the total system performance (including the human component).

The control room in its entirety should be designed with a high level of comfort. Low comfort levels result in operating staff becoming unnecessarily fatigued more quickly, which will lead to a reduction in monitoring ability. Sometimes the opposite argument is put forward: if the control room is too comfortable, the vigilance of staff is reduced. This is simply not correct. A comfortable and cosy environment increases the possibility of operators remaining awake and alert. It is also important to create a varied information situation in the control room, for example, easy access to radio and television. However, this secondary source of information should be designed and located so that it does not in any way disturb the primary work tasks, especially during periods of high work stress.

Comfortable, adjustable chairs and an ergonomic layout of the control room furniture are very important issues, and a great deal of consideration should be given to lighting conditions. Visual display units place great demands upon the design and location of light fittings. In order to avoid serious visual problems it is recommended to seek advice from a qualified lighting specialist. Glare can very seriously impede visual performance. Considering the unique conditions found in control rooms (for example, 24 hours of continual operations), it may even be desirable to use full spectral lighting, that is, lighting that provides 'daylight' conditions.

When planning a new control system it is important to create situations where, in an unbiased way, alternative forms of automation can be tested. The planners of the system must realise that their way of thinking and thus their concept of the system is often very different from that of the process operator. For example, a solution that seems obvious and correct to a systems designer may be quite alien to a control room operator. Therefore, when planning new systems, new methods must be found in order to bridge the differences in conceptual thinking between designers and operators. The literature, including this handbook, provides us with much information about the operators' capabilities and limitations. There is also considerable knowledge about the capabilities and limitations of the different senses. Further, there are data about perception, cognition, and decision-making processes, and also about learning, training, and education. By consulting each of these areas, operators can attain a suitable level of skill for control room work.

The trends that are predominant today with regard to the design of control rooms and control systems are not synonymous with a development that creates an optimal system from a human/machine point of view. An increased understanding of humans can result in new products and new systems when developing the process control systems of tomorrow. This will increase safety significantly and enable better production, which is also in harmony with the environment.

Those manufacturers who systematically invest in development of systems and equipment that place humans in the centre have found a business concept that can make the organisation a leader in its field. Over the past hundred years we have learnt that technology more and more balances on the edge of what is possible and desirable from the human point of view. A more humane technology is not only a human demand; it is also a prerequisite if we are to reap further rewards with regard to improvements in productivity and quality from technology usage.

14.5   SUMMARY OF A NEW PERSPECTIVE

Why is it necessary to discuss the role of people in control rooms? For many of us it has been obvious for decades that we need to create a harmony between technology and the people involved in steering, controlling, and managing the technology. Countless accidents with very dramatic and severe consequences have been blamed on 'the human factor'. The human and environmental tragedies of Bhopal, Brent Spar, Chernobyl, the *Exxon Valdez*, MinaMata, and Three Mile Island show what can go wrong when humans engage with machines. However, very early we learned and understood that the operator was not to blame. The reasons for the catastrophes were defined as the *lack of compatibility* between people and technology. In essence, the interface between technology and people did not match. Today we talk about a lack of usability and ergonomic considerations. Earlier we used words like 'human factors' and 'human engineering'. If some human agents need to be blamed it is the designer and the purchaser of the systems for overlooking the critical importance of technology/human factors. Most readers of this book are aware of these factors, but in some parts of the world this remains a little understood area of knowledge. A 'nice-looking' control room is more prized as a place to show and impress visitors. This seems to imply that form is of greater importance than functionality and use. This is obviously not so. A careful design of the interrelationship between the operators and the control systems and its controlled processes is essential for a safe and optimal operation.

Every 15 to 20 years we see a paradigm shift in industrial and technological processes. The introduction of new technologies into organisational processes requires organisations (and their members) to 'change paradigms' about how they work and behave (Clarke and Clegg, 1998). The inertia in this process of change is related to many factors, but probably the most important is the need for return on investment. Rapid developments in technology (including information technology [IT]) create an urgent need for organisations to develop new skills, competencies, and knowledge (Ivergård, 2000; Paulsson, Ivergård, and Hunt, 2005). The three elements of technology are: as a driver of organisational transformation, as an enabler to deliver that transformation, and as a change agent of the organisation to include new skills and competencies. For many (if not all) organisations, a critical issue is not in keeping up to date with the development of technology *per se* but in developing competencies to enable the organisation to gain full advantage of the technology.

The full potential of technology will only be realised if the organisation succeeds in adopting new business processes and working routines into the organisation as new ways of doing things (McKenney, 1995; Clarke and Clegg, 1998). Thus it becomes critical for organisations to develop new competencies in employees who use the new technology in their work. Introducing technologies into organisational processes requires organisations (and their members) to 'change paradigms' about how they work and behave (Clarke and Clegg, 1998). This includes learning the skills needed to utilise new technologies in the workplace (Edmondson et al., 2003). Technology is thus a key driver for workplace learning as employees need to develop skills for managing and operating the newly-adopted technology (Pisano, 1994; Paulsson et al., 2004). Learning the skills to perform using the new technology is critical if the technology is to be used for optimum benefit.

However, primarily, the new technology in itself has to be adapted to the existing competence and knowledge infrastructure of the organisation. This is the ergonomic approach to technological change. The importance of this symbiotic relationship between technology and people in organisations is discussed by Ivergård (2000). Technology is normally not an aim in itself but a means to achieve other aims, for example, to improve efficiency of learning or to reduce cost of learning. Inherent in learning at work are many possibilities of integrating learning technologies as a part of the control system of industrial and administrative processes. As such, technology has been a key driver for learning at work (Pisano, 1994; Edmonson et al., 2003). Intelligent technologies need to learn and adapt to the skill, knowledge, habits, and culture of the operators. Artificial intelligence (AI) is a good example. AI is a new and potentially very fruitful area of application. However, it is obvious that e-learning and other forms of learning technologies can never—or, at least, very rarely—'stand alone' for learning process. Rather, it has to be combined with other methods to create a holistic process of learning.

To build a good society we need to create harmony and balance between people and between people and technology. To create a good production environment in balance with nature we also need a good harmony and balance between people and technology. It is important for decision makers to understand these points. It is necessary to define the problem, to recognise the benefits of worker participation and motivation to design and create, and to take appropriate action to engage these.

In less than a generation, we will probably see different 'new generation' control centres. One technological development is likely to be a vertical high-tech AI-supported centre automatically capable of optimising the processes according to preset control variables. The main task of the operator will be to continually update and upgrade the computer software and its survival systems.

The second might be an organic control system continuously supporting the operator's tacit knowledge base and mental models of the process. It will be a balanced symbiosis involving a harmonious relationship between the human operator and the technological system. Here the operator uses all senses to support learning and systems knowledge. This will in turn be used to upgrade the systems software and hardware. In both cases the system will include corporate social responsibility (CSR) and environmental and business considerations.

In this handbook we have made the case, from various perspectives, of the need for usability and ergonomic design to be incorporated into control room design and use. With the increasing importance of control rooms as key components in the operations and management of businesses and governments, the competencies of the operators take on a new leading role. In this new and evolving situation the operator is not only handling a limited industrial process. The work in the control room is becoming a focal point for the success of many different types of organisations in, for example, business, trading, energy, environment, and government. From many of the latest control centres we have been investigating, it is clear that the levels of competence and necessary education of operators has increased. It is also clear that control centre tasks cover a larger perspective than in the past. As we have discussed elsewhere in this handbook, creative development tasks have become integrated into control centre work.

This trend is likely to continue and to expand. The advanced control centre of the future will also include top-level experts and senior executive decision makers. This will most likely be the case in government organisations as well as in business-orientated industries. This might also be the case in international organisations, such as the United Nations, the World Bank, and the Asian Development Bank. It is easy to see the advantage for top-level executives such as managing directors and presidents of large public and private organisations to have direct access to extensive databases and real-time process information presented on very large-scale wall displays.

Furthermore, the availability of simulation programs of macro-perspectives of business relations and business actions will help decision makers improve the quality of their decisions and predictions. This future type of control centre is also likely to create its own architecture—both physical and functional. The physical architecture of the control centre of the future will be a very challenging task for architects and designers.

REFERENCES AND FURTHER READING

Clarke, Thomas, and Clegg, Stewart. (1998). *Changing Paradigms: The Transformation of Management Knowledge in the 21st Century.* London: Harper Collins Business Books.

Edmondson, A., Winslow, A., Bohmer, R., and Pisano, G. (2003). Learning How and Learning What: Effects of Tacit and Codified Knowledge on Performance Improvement Following Technology Adoption. *Decision Sciences,* 43, Spring, 197–223.

Ekvall, Göran. (1991). Change-Centred Leaders: Empirical Evidence of a Third Dimension in Leadership. *Leadership and Organizational Development Journal* 12, 6, 18–24.

Florida, Richard. (2004). America's Looming Creativity Crisis. *Harvard Business Review,* 82, 10, 122–136.

Florida, Richard. (2005). *The Flight of the Creative Class: The New Global Competition for Talent.* New York: Harper Collins Books.

Hunter, Samuel T., Bedell, Katarina E., and Mumford, Michael D. (2007). Climate for Creativity: A Quantitative Review. *Creativity Research Journal* 19, 1, 69–90.

Isaksen, Scott, G. (2007). The Climate for Transformation: Lessons for Leaders. *Creativity and Innovation Management* 16, 1 (March), 3–15.

Isaksen, Scott G., Lauer, Kenneth J., Ekvall, Göran, and Britz, Alexander. (2001). Perceptions of the Best and Worst Climates for Creativity: Preliminary Validation Evidence for the Situational Outlook Questionnaire. *Creativity Research Journal* 13, 2, 171–84.

Isaksen, Scott G., and Lauer, Kenneth. (2002). The Climate for Creativity and Change in Teams. *Creativity and Innovation Management* 11, 1 (March), 74–86.

Ivergård, Toni. (1982). *Information Ergonomics.* Lund, Sweden: Studentlitteratur.

Ivergård, Toni. (2000). An Ergonomics Approach to Work in the Next Millennium in an IT World. *Behaviour and Information Technology,* 19, 2, May–June.

Ivergård, Toni, and Hunt, Brian. (2007). A High Ceiling Environment for Learning and Creativity. Paper presented at the 4th ASIALICS Conference. Kuala Lumpur, Malaysia, 22nd–24th July.

Lancaster, John (ed.). (2005). *Engineering Catastrophes: Cause and Effects of Major Accidents* (3rd ed.). Boca Raton, FL: CRC Press.

McKenney, J. (1995). *Waves of Change: Business Evolution through Information Technology.* Boston: Harvard Business School Press.

Murrell, K.F.H. (1967). Why Ergonomics? *Occupational Psychology* 41, 17–24.

Paulsson, Katarina, Ivergård, Toni, and Hunt, Brian. (2005). Learning at Work: Competence Development or Competence-Stress. *Applied Ergonomics,* 36, 135–144.

Peters, George A., and Peters, Barbara J. (2006). *Human Error: Causes and Control.* Boca Raton, FL: CRC Press.

Pheasant, Stephen, and Haslegrave, Christine M. (2005). *Bodyspace: Anthropometry, Ergonomics and the Design of Work.* Boca Raton, FL: CRC Press.

Pisano, G. (1994). Knowledge, Integration, and the Locus of Learning: An Empirical Analysis of Process Development. *Strategic Management Journal,* 15, Winter, 85–100.

Sackman, H. (1967). *Computers, System Science and Evolving Society: The Challenge of Man-Machine Digital Systems.* New York: John Wiley & Sons.

Vilkinas, Tricia, and Cartan, Greg (2001). The Behavioural Control Room for Managers: The Integrator Role. *Leadership and Organizational Development Journal* 22, 4, 175–85.

Waller, Mary J., Gupta, Naina, and Giambatista, Robert C. (2004). Effects of Adaptive Behaviors and Shared Mental Models on Control Crew Performance. *Management Science* 50, 11 (November), 1534–44.

Wu, Ji-Tsung Ben, and Marakas, George M. (2006). The Impact of Operational User Participation on Perceived System Implementation Success: An Empirical Investigation. *Journal of Computer Information Systems* 46, 5, 127–40.

Index

Bold locators indicate material in figures and tables.

A

ABB; *See* Asea Brown Bouverie
Absolute discrimination, 76, 77, 267
Acoustic environment
    music, 246, 286
    noise; *See* Noise
    signals; *See* Sound signals
    speech; *See* Speech
Action research (AR), 37–38, 302, 327
Active vision, 63–64
Actual value, 20, 22
Advanced Technology to Optimise Maritime
        Operational Safety (ATOMOS)
    focus of, 229
    on health hazards, 258
    HFE for SCCs, 233
    on safety, 259
    on ship's bridge design, 235–236, **236**,
        238–239
    on survivability, 259
Age factors, 282–283
AI; *See* Artificial intelligence
Aiding mechanisms, 22–23, **24**, 25
Air conditioning
    infrasound from, 198
    for large screen system, 126–127
    noise from, 178, 194, 195
    planning for, 177
    speech and, 195
Air pollution, 211, 223, 224, 292
Air speed
    guidelines for, 183–184
    human heat production and, **180–182**
    noise from, 194, 197
    temperature and, 179, **184**
    units of, 179
AIS; *See* Automatic identification systems
Alarms
    acknowledgement of, 254, 256–258
    audible, 57, 231, 256, 258
    at Billerud, 209

    in engine control room, 248–249, 253–258,
        255
    equipment for, 58–60
    IMO on, 256–258
    on instruments, 57, 255
    large screen systems and, 123
    layout and, **175**
    presentation of
        chronological, 60, 257
        graphic vs. linear display of, 101–102
        hierarchy, 57, 60
        IMO on, 256–258
        random order, 60
        tabular form with text, 60, 256
        traditional, 59
    reaction to, 280
    in SCADA, 130
    on ship's bridges, 231, 255–258
    SOLAS on, **238–239**, **249**
    staffing and, 37
    on VDUs, 164, 257, **257**
    visual, 164, 257–258, **257**
Alertness level, 284–288, **285**, **287**
Ambient light
    colour and, **270**
    contrast and, 98, 103
    controlling, 98
    on controls, 188, 189
    definition of, 97
    design of, 211, 350
    edge lines of, 192–193
    eye adaptation level and, 98, 269–270
    front projection and, 102–103
    guidelines for, 185–193
    on instruments, 189–191
    intensity of, 184–185, 269
    LCDs and, 111
    mathematical modeling of, **186–187**
    onboard ships, 237–240, **241**
    performance and, 185, 269
    plasma displays and, 118
    rear projection and, 104, 110
    reflection factors and, 185–189, **188**, 192–193
    shading of, 189, 192
    sources of, 97, **98**
    for standing work, **165**

T - #0347 - 071024 - C7 - 234/156/18 - PB - 9780367386733 - Gloss Lamination